THE UNIVERSE
Its Beginning and End

THE UNIVERSE

Its Beginning and End

LLOYD MOTZ

ILLUSTRATED WITH PHOTOGRAPHS AND DIAGRAMS

CHARLES SCRIBNER'S SONS / NEW YORK

Library of Congress Cataloging in Publication Data
Motz, Lloyd, 1909–
    The universe—its beginning and end.

    Bibliography: pp. 331–332
    Includes index.
    1. Cosmology. I. Title.
QB981.M87   523.1   75–6635
ISBN 0–684–14239–2

1 3 5 7 9 11 13 15 17 19 C/C 20 18 16 14 12 10 8 6 4 2

PRINTED IN THE UNITED STATES OF AMERICA

Frontispiece. Nebulosity in Monoceros, in the south outer region of NGC2264, shows turbulence in a gas and dust cloud where stars are being formed.

To Minnie

CONTENTS

ILLUSTRATIONS

PREFACE

The idea for this book originated in the fall of 1973 when I was writing an article on cosmic chemistry for *Chemical Engineering*. As I developed the material for the article, it occurred to me that many of the topics, if properly expanded and written in nontechnical language, would be of considerable interest to the general public. The editors at Charles Scribner's Sons agreed with me, and so this book was conceived and written, with the emphasis placed on the origin, the evolution, and the ultimate collapse of the universe.

The most difficult problem that faces the writer of a book such as this is in the choice of material; astronomical knowledge is so vast and the topics are all so interesting that the tendency is to introduce too much rather than too little. To avoid this, I have tried to use only material directly related to or important for an understanding of the main subject, but even so, reducing it to manageable length was not easy. Because the book is designed to appeal to the general reader, I also had to decide just how far to go in discussing advanced theories. The two great twentieth-century physical theories—the theory of relativity and the quantum theory—have had so great an impact on all phases of science and on our understanding of man's place that it is impossible to write intelligently on the origin and evolution of the universe without treating them, even if only briefly. Wherever required, the content of these theories is woven into the fabric of the text in such a way that phenomena that would otherwise remain obscure are clarified.

Classical science, as it evolved out of the Newtonian ideas of absolute space and absolute time, was based on the assumption that an event is described in the same way by all observers and has the same meaning for every observer. The intelligent observer, according to this point of view, played only the role of a recorder of events; it was unnecessary to take his presence into account in formulating the laws of nature. But according to relativity theory and the quantum theory, the observer cannot be neglected in formulating natural laws, and these laws can be properly understood and interpreted only if the role of the observer and his frame of reference are properly delineated. If one accepts this point of view, one can hardly consider a book on the evolution of the universe complete if it does not include a section on the emergence of intelligent life. Scientists have always concerned themselves with man's place in the universe, but never before has this been so relevant to an understanding of the universe. The fate of man seems as deeply hidden from our view as ever, but insofar as it is bound up with that of the universe there is reason to hope that in solving the problem of the evolution of the universe we shall also solve the problem of man's destiny. In the last chapter I discuss these questions and indicate what contemporary cosmologists have to say about them.

Grateful acknowledgment is made to Hale Observatories (Mt. Wilson and Palomar) for permission to use the photographs which appear in this book.

Some readers may need an explanation of the letters NGC and M, which appear in conjunction with numbers in the text and the photograph captions. NGC stands for New General Catalogue (1888) and M for the Messier Catalogue of Nebulae and Star Clusters (1787); both are standard works used by astronomers.

THE UNIVERSE
Its Beginning and End

1 On Structures and Forces

Diversity and Order from Chaos

By far the most striking and remarkable feature of the world, and, indeed, of the whole universe, is the great diversity and variety of the objects that compose it. All of creation, from the vast clusters of galaxies, rushing away from each other as the universe expands, down to the molecules and atoms of the human mind and body, which give rise to all of man's thoughts and actions, reveals a bewildering and ever-changing hierarchy of structures and forms. But in all of this variety and endless change, with no single thing abiding and all things flowing, there are patterns that abide and forms that exist for millennia everywhere in space and time. The delicate rose petal emerging from its bud each summer, the ocean waves continually pounding the continental shores, the moon waxing and waning through its phases month after month, the planets moving repeatedly through the same stellar constellations, the stellar constellations themselves, and the almost exact mathematical symmetry of the spiral nebulas are examples of forms and structures that recur time after time throughout the universe, as though there were identical molds throughout space into which the matter of the universe had been poured.

But even more remarkable than the existence and persistence of such a diversity of forms is the manner in which this diversity evolved in the course of 13 billion years from a universe that consisted initially of a chaotic, but homogeneous, mixture of only three things: radiation, hydrogen, and helium. Not a trace of all the present wonders of the universe could have been discerned in its initial, chaotic state, in which not even the simplest chemical reactions could occur, let alone the incredibly complex ones required to support even the simplest form of life. How, then, did all these wonders—and, in particular, life—arise from such a sterile, chaotic medium?

Since the world and the living organisms on it consist not only of hydrogen but also of carbon, nitrogen, oxygen, phosphorus, sulfur, iron, and other heavy elements that were not present in this primordial state of the universe, all the heavy elements, from carbon to uranium, had to be synthesized from the original hydrogen and helium before the world could come into being. There were at least two distinct stages in the history of the molecules and atoms that now constitute all the bodies in the solar system.

The first stage, which probably lasted for about 6 billion years and during which the oldest stars were formed, led to the gradual transformation inside these stars of about 3 or 4 percent of the original hydrogen and helium into the heavier elements without which life would be impossible. The second stage, which also lasted for about 6 billion years, led to the formation of a second generation of stars like the sun and their retinues of planets.

A relatively short third stage of the order of a few hundred million years led to the formation of complex organic molecules that began to replicate at a certain stage of complexity and were thus the precursors of the first living cells. Just why and how these first complex organic molecules began to replicate is a mystery, but it is known that complex organic molecules (for example, the amino acids, which are necessary for life) are built up quite naturally in interstellar space when conditions are proper and the necessary chemical elements are present. Direct evidence for this can be found on the surfaces of meteorites that come from outer space and strike the earth.

The emergence of life from inorganic matter at the end of this short third stage appears to be as natural and inevitable a phenomenon everywhere in the universe as the formation of the galaxies, the stars, the chemical elements, and the planetary systems. This in itself is the most amazing phenomenon of all in the universe, since nothing that portends life can be detected in the behavior or the chemical properties of any of the chemical elements. More remarkable still is the ultimate evolution of a highly intelligent form of life that can contemplate nature and discover the laws that govern it and govern life itself; a small part of nature laboriously studies all of it and discovers the principles on which the whole of nature is based.

The Force Concept

The transformation of primordial hydrogen and helium into heavy elements requires the very high temperatures and pressures that can be found only near the centers of stars. The question of how these first stars were formed from a chaotic, gaseous medium that was expanding rapidly in all directions at an incredible speed can be answered in detail only after a consideration of the more general question of why such structures as galaxies, stars, solar systems, planets, rocks, molecules, atoms, and nuclei exist at all. A universe without such structures could theoretically exist, consisting of only such basic particles as protons and electrons rushing randomly about, without any fixed patterns in their movements or restraints on their motions. Such a universe would be devoid of structures that would arise if the motions of the individual particles were restrained in some way. The existence of the various complex structures in the universe is, then, clear evidence that restraints on, or forces between, particles exist and that these forces give rise to the structures. The existence of structures therefore means the existence of forces and vice versa. The great variety of structures that are everywhere present might be thought to imply that there are many different forces in nature to give rise to these structures, but this is not so. The remarkable fact is that there are only four known forces in nature, and they are sufficient to account for the great variety of things. A brief discussion of the general nature of a force is necessary here, since the force concept plays such an important role in an understanding of the structures in the universe.

All our actions and life processes are related to, and governed by, forces, whether we are directly aware of them or not. Whether we sit down, rise from a chair, walk across the room, raise our arms, or wink our eyes, we are exerting forces on something or having forces exerted on us. In such instances we are directly aware of the forces, because they are associated with a deliberate muscular effort on our part. Many internal bodily functions—flow of blood, breathing, digestion—are also caused by forces, but since, under ordinary circumstances, we are not conscious of the muscular efforts that generate these forces, we are not aware of the forces themselves.

Force may be loosely defined as a push or pull exerted by one body, or collection of bodies, on another. We are immediately aware of forces acting on our own bodies because of muscular responses to such forces. In fact, human muscles are so sensitive and responsive to forces that it takes little conscious effort for us to make the muscles exert just the right amount of force to perform all kinds of incredible feats. Life would be intolerable—and perhaps impossible—if, for example, a person had to estimate consciously the forces required for such acts as opening a door, walking at a fixed pace, or picking up a piece of bread and had to direct the muscles accordingly. We do these things automatically because our muscles quickly learn to estimate forces and respond by just the right amount to perform each task easily. Very remarka-

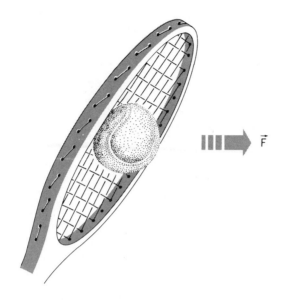

A tennis racquet exerting a force on a tennis ball.

ble examples of this are found in the performances of top acrobats, athletes, and virtuoso musicians, whose muscles are so attuned to precise actions that some of their feats border on the miraculous. Both the champion tennis player deftly stroking the ball so that it floats over the net for a perfect drop shot and the concert pianist striking each key with just the right amount of force to give his performance its quality of greatness do so because they have acquired almost perfect muscular response.

We are directly aware of external forces acting on our own bodies because of our muscular responses to these forces, but how do we know when a force is acting on some other body? If we see this other body being pulled or pushed by means of a rope or a rod in contact with it, we may safely surmise that a force is acting on it in the direction of the rope or the rod. In general, if two bodies are in contact, each of the bodies is assumed to be exerting a force on the other. But how does one know that a force is or is not acting on an object that

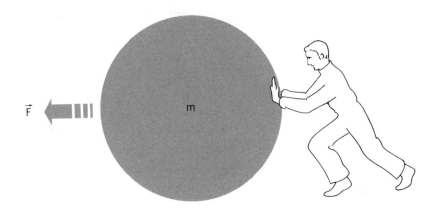

A force $\vec{F}$ imparts an acceleration to a body of mass m.

is in contact with no other body? In that case a study of its motion must be made before an answer can be found. If the speed or direction in which an object is moving is changing, a force is acting on it. If the body is moving with constant speed in a straight line or is at rest, then there is no net, or unbalanced, force acting on it. In other words, the action of an unbalanced force on a body shows up as a change in the velocity of the body.

Some specific examples will illustrate this and show what is meant by a net force. If an automobile is speeding up along a straight road, a net force is pushing the automobile in the direction of its motion along the road. It is actually the earth, via friction, that is exerting the force on the automobile and thus accelerating it. If an automobile is slowing down as it moves along a straight road, there is a net force—again the friction of the earth—pushing the automobile in a direction opposite to that of its velocity. If an automobile is standing still or moving along a straight road with constant speed, there is no net force acting on it in any direction. It is true that there are various bodies, such as the air or the earth, acting, and thus exerting forces, on this automobile, but all of these forces balance each other, and thus cancel out, as long as the automobile's speed and direction of motion do not change.

Consider now a body in a vacuum that starts from rest at a certain height above the surface of the earth and is allowed to fall freely. Since its speed increases at a constant rate as it falls to the ground, it follows that there is an unbalanced downward force acting on the body even though there is no other body in contact with it pulling it to the ground. Another example of the action of this same force is found in the motion of a comet around the sun or of an artificial satellite around the earth. This force changes the direction of motion as well as the speed of the artificial satellite every moment, thus causing the satellite to orbit the earth. The earth and the planets are propelled in their orbits around the sun by the same kind of force that causes bodies to fall to the earth with increasing speed and causes satellites to circle the earth.

It is apparent, then, why, in a general way, forces give rise to structures. If there were no forces in nature, the particles that form any structure would move off in straight lines in all directions instead of remaining together as they are compelled to do by the forces that bind them to each other.

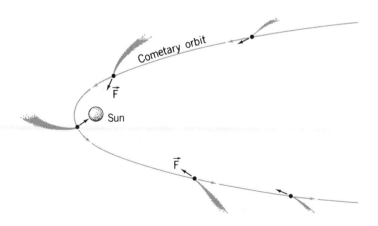

The force of gravity from the sun accelerates a comet around the sun. Note the change in the direction of the comet's motion.

The force required to change the velocity of a body by a given amount in a given time is not the same for all bodies. A baseball catcher applies very little force when he catches a pitched ball and reduces its speed from about 100 miles an hour to zero in less than a second, but the same catcher would have to exert an enormous force if he were to reduce the speed of an automobile from 100 miles an hour to zero in less than a second. Obviously, then, in addition to the rate at which the velocity of a body is changing, some other physical element associated with the body enters into the magnitude of the force acting on the body. This physical element is called the mass of the body.

The more massive a body is, the greater is the force required to change its velocity by a given amount in a given time. It is customary to think of mass as "the quantity of matter in a body," which is the definition of *mass* given in some physics books. But this does not really say anything about the nature of mass, because mass and the quantity of matter are different expressions for the same thing. But the relationship between force and mass does say something about the nature of mass: it is that property of an object which resists changes in the velocity of the body and thus accounts for the inertia of a body. The greater the inertia of a body, the more massive it is.

The inertial property of mass can be used to compare the masses of different bodies. For example, if two bodies are moving with the same speed in a straight line and twice the force is required to bring one of them to rest as the other in a given time, that body has twice as much mass as the other.

The British mathematician Sir Isaac Newton (1642–1727), who discovered the relationship between force and motion described here, completed his discussion of the laws of motion by stating what is now known as the law of action and reaction. This remarkable law says that if body A exerts a force on body B, then B exerts an exactly equal force in the opposite direction on A. Thus, for every force in nature acting on a body, there is an equal and opposite force acting on some other body, so that forces in the universe come in equal and opposite pairs. In other words, when all the forces in the universe are added up, the sum is zero. There is no net force acting on the universe itself.

There are many interesting examples of this law of action and reaction that, if properly understood, clarify many otherwise puzzling phenomena. Man can walk and an automobile can ride along only because of this law. Man walks by first pushing backward against the earth with his feet, one foot at a time, which causes the earth to react against his feet and to push him forward with an exactly equal force. In the same way, the automobile is pushed forward by the earth when its tires push the earth backward. Another example of this law is that a freely falling body pulls on the earth with a force that is exactly equal, but opposite, to the force with which the earth pulls the falling body downward. On a much larger scale, the earth, as it revolves around the sun, pulls on the sun with a force that is exactly equal to the force with which the sun pulls on the earth. The same is true of each planet as it moves in its orbit around the sun. The only reason that pulls exerted by the earth and other planets on the sun do not accelerate the sun by any appreciable amount is that the sun is so very massive that its response to these forces is negligible.

Another interesting example of action and reaction is that of the baseball pitcher and the ball he is throwing. The ball pushes against his arm with a force exactly equal to the force with which his arm pushes the ball forward. If the pitcher were not firmly rooted to the ground—for example, if he were on skates or on perfectly smooth ice—he would move backward in response to the push of the ball as he threw it forward.

If a force acts on a body in the same direction as that in which the body is moving, the force increases the speed of the body without changing the body's direction; but if a force acts in the opposite direction, it slows the body down. If a force always acts at right angles to the direction of motion of a body, the direction of motion of the body changes continuously but its speed remains unaltered; the body thus moves in a circle with constant speed. In general, forces act on bodies in arbitrary directions, so that both their speeds and directions of motion change. The directional character of forces places them in a category of physical entities that are collectively called vectors. *Vector* is the name that physicists apply to an entity that has both magnitude and direction. The velocity of a body is another example of a vector.

The Four Known Forces of Nature

GRAVITY

Newton was the first to surmise that every particle of matter in the universe interacts with every other particle via the force of gravity and that not only the motion of a falling apple but also the motions of the planets around the sun and of the moon around the earth can be understood in terms of this force. Newton's great achievement was his discovery of the physical characteristics of the force of gravity, which he formulated as a mathematical law for two particles pulling upon each other.

In general, two different physical characteristics contribute to any force between two interacting particles. One is a geometrical quantity that is related in some way to the distance between the two particles, and the other is a physical quality intrinsic to each particle. The presence of the distance factor in the force law is fairly easy to grasp, for it is clear that the greater the separation between two particles, the weaker is their influence on each other. In the case of the force of gravity Newton showed that doubling the distance between two particles reduces the gravitational force by a factor of 4, tripling the distance reduces the force by a factor of 9, quadrupling the distance reduces the force by a factor of 16, and so on. The force of gravity is said to fall off as the square of the distance between the two particles. This is the inverse-square law. It should be noted that although the gravitational force between the particles gets weaker and weaker as the distance between them increases, it never vanishes, no matter how far apart the two particles may be.

The intrinsic physical quality of each particle that enters into Newton's formula for the gravitational force is the mass of each particle. The more massive the particles are, the greater the gravitational force between them. The exact statement of this fact is that the gravitational force between two particles depends directly on the product of the masses of the two particles. This means that if, for example, the mass of one of the particles is doubled and the mass of the other is tripled with no change in the distance between them, the gravitational force between them becomes six times greater.

Daily experience shows that the gravitational force between two ordinary bodies, even if they are very close to each other, is extremely weak; the human body detects no gravitational pull toward the most massive buildings even when it passes very close to them. We do, however, detect the gravitational force exerted on us by the earth, because the earth is so very massive. This force is referred to as weight and is expressed in pounds; one pound is the force that the muscles exert when one lifts up about four apples of average size. In calculations involving forces, scientists use a unit of force called the dyne, which is about 450,000 times smaller than the pound.

Since the gravitational pull of the earth on any object is always directed toward the center of the earth, as though the entire mass of the earth were concentrated at its center, the weight of a person, no matter where he is on the surface of the earth, is always directed toward the center of the earth. The failure to understand this basic gravitational phenomenon led many people before the time of the great explorers to believe that the earth had to be flat to enable people to stand on it. The weight of an object on the earth, then, is nothing more than the gravitational pull of the earth on that object. If the object is placed on the moon, its weight is different because the mass and size of the moon are smaller than those of the earth. In fact, an object on the moon is only one-sixth as heavy as it is on the earth.

Although the gravitational pull between the ordinary objects surrounding man is extremely small—indeed, quite negligible—it is enormous between massive bodies like the sun and the planets. Thus, even though the earth is 93 million miles away from the sun, the gravitational pull of the sun on the earth (and of the earth on the sun) is about 6 billion trillion (6 followed by 21 zeros) pounds. Weak as the gravitational force is on the surface of a body like the earth under ordinary circumstances, there are situations in which it can become great enough to overcome every other force. To see how this comes about, consider what would happen if the earth were suddenly reduced to half its size with no change in its mass. All objects on the earth's surface would then be half as far from the center of the earth, and hence would weigh four times as much, because the gravitational force varies as the inverse square of the distance from the center of attraction—in this case, the center of the earth. If the earth were suddenly reduced to a quarter of its present size, all things on the earth would become 16 times heavier, and so on. If a star like the sun were to contract, a point would be reached where all objects on its surface would be subjected to so intense a gravitational force that none of them could withstand it.

One very important characteristic of the gravitational force accounts for weightlessness, which plays such an important role in the lives of astronauts when they are in a space vehicle. The Italian scientist Galileo (1564–1642) was the first to surmise that all bodies in a vacuum, regardless of their mass, fall to the earth with exactly the same speed. A heavy boulder and a feather placed in a vacuum together above the

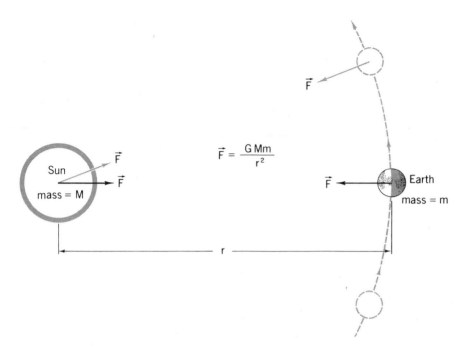

Newton's law of gravitational force and action and reaction. The pull of the sun on the earth is exactly equal and opposite to the pull of the earth on the sun. Its magnitude is GMm/r^2, where G is the gravitational constant and r is the distance between the center of the sun and the center of the earth. The pull of the sun on the earth keeps the earth moving in an elliptical but very nearly circular orbit around the sun. The sun also moves, but very slowly and in a very tiny orbit.

earth's surface and allowed to fall, do so equally fast and reach the earth at the same moment. To see how this relates to weightlessness, note that when someone weighs himself, he stands on a scale that is fixed on the ground, so that his feet push against it. Suppose now that while he is standing on the scale, the floor under the scale is suddenly removed so that it is free to fall. Since the scale falls as fast as he does, his feet can no longer push against it; consequently, the scale registers no weight and the man has thus become weightless. This does not mean that there is no force of gravity acting on him; it simply

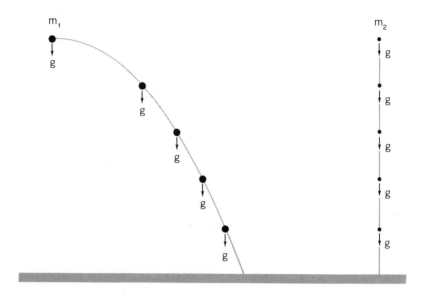

Two bodies of different weight (different mass) allowed to fall freely in a vacuum from the same height above the ground experience the same acceleration; if they are released or thrown horizontally at the same moment, they reach the ground at the same moment.

means that he is in free fall under the action of gravity. All the objects in a spacecraft become weightless when the spacecraft is falling freely in a vacuum, because everything in the craft falls as fast as the craft does. The great importance of this phenomenon was not realized until the German theoretical physicist Albert Einstein (1879–1955) made it the basis of his general theory of relativity.

Gravity possesses a number of other interesting features. It cannot be deflected nor can any object be shielded against it; there is no such thing as antigravity material, or material that can cancel the gravitational force. Moreover, the force of gravity of two objects on each other is not altered by the presence of a third body; two bodies interact gravitationally with each other as though no other bodies were present. Another important feature of the gravitational force, which was first predicted by Einstein in his theory of relativity, is that it acts on energy (light and radiation) as well as on matter and that it affects space and time; time slows down as a result of the gravitational force, and space is warped, which means that the geometry which describes it is non-euclidean.

Consider now the gravitational force emanating from a body like the earth. This force extends outward in all directions, becoming weaker and weaker as one moves away from the earth, but never vanishing. Gravity permeates all of space, so that it may be referred to as a field of force generated by a given body. Instead of speaking of the gravitational interaction of two bodies, it is convenient to speak of the interaction of either one of the bodies with the gravitational·

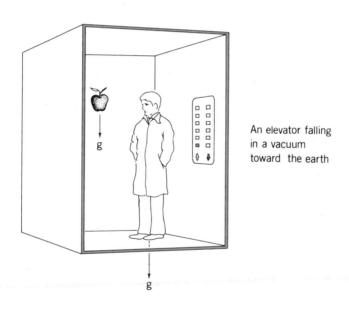

An elevator falling
in a vacuum
toward the earth

All objects in a freely falling elevator are weightless, because they all fall with the same acceleration g. This is also true of astronauts circling the earth.

field generated by the other. Thus, one may say that the earth revolves around the sun in its prescribed orbit because it is compelled to do so by the gravitational field of the sun.

Gravitational force is considered the prime force in the universe not because it is strong—it is in fact quite weak—but because it is all-pervasive and because it had to organize the original chaotic matter of the universe into such vast structures as stars, planets, solar systems, and galaxies long before other forces could come into play. In general, weak forces give rise to large structures, and strong forces to small structures; the stronger the force, the smaller the structure. It is precisely because the gravitational force is so weak that its influence is all-pervasive and that it gives rise to the very largest structures in the universe. And it is also because of its weakness that this force, as far as is known, plays no measurable role in the structure of matter. It is quite easy to lift a small piece of steel wire off the ground but very difficult to pull it apart, which shows that the gravitational pull of the entire earth on the wire is negligible compared to the internal forces that keep the atoms of the wire together. These internal forces, being very strong, are quite different from the gravitational force. They are an example of one of the many different forms in which the electromagnetic force reveals itself.

ELECTROMAGNETISM

The various forms of the electromagnetic force seem so different from each other that it took scientists a long time to realize that they are all intimately related and are basically the same force. Friction, the push of the wind against bodies, the resistance of a hard surface to the touch, and the impact of the head of a golf club driving the ball are a few other examples of this amazing force.

Whereas the gravitational force derives from the mass of a body, and hence is always present between any two bodies, the electromagnetic force stems from the presence of an electric charge on a body and is therefore present between two bodies only when both bodies are electrically charged. Under ordinary circumstances, the objects that we live with are electrically neutral, or uncharged, so that there are no electrical forces between them. That bodies can be charged electrically by rubbing them against each other was discovered by the ancient Greeks. They found that pieces of amber could be charged by rubbing them with fur and that the charged amber then attracted bits of matter to it. We can charge our own bodies by scuffing the soles of our shoes across a rug; when we then touch some other object—the radiator in a room, for example—the electric charge leaves us very quickly in the form of a spark. When storm clouds rub against each other, they become electrically charged; and when they discharge themselves to the earth, lightning is produced. All of these phenomena are the result of the electromagnetic force, and all attest to its great strength.

There are two kinds of electric charges in nature—the positive and the negative—and two charges will attract or repel each other depending on whether they are unlike or like charges. Thus the effect of one kind of charge can be offset by the presence of the other kind. If equal quantities of positive and negative electric charges are in or on a body, they offset each other and the body behaves as though it had no electric charge on it at all. When a piece of amber is rubbed with some fur, some of the negative charge is rubbed off the fur onto the amber, leaving a slight excess of negative charge on the amber and an equal excess of positive charge on the fur. This excess charge on the amber attracts objects like small pieces of paper to it. An electric current is obtained when one end of a metal wire is attached to a body having an excess of negative charge (it is negatively charged) and the other end to a body having an excess of positive charge (it is positively charged). The positive charge pulls the negative charge through the wire, thus creating a flow of electric charge, which is called an electric current. The electric batteries in automobiles are devices that separate positive and negative charges chemically and concentrate these charges onto two different metallic rods called the positive and negative electrodes, or poles, of the battery.

Magnetic forces come into play when an electric current is flowing—that is, when electric charges are in motion. In 1820 the Danish physicist Hans Christian Oersted (1777–1851) discovered that a magnet pulls on a wire in which an electric current is flowing and that the wire, in turn, pulls on the magnet. In fact, two wires in which electric currents are flowing pull upon each other magnetically. This phenomenon is the basis of the electric motor. In general, when electric charges are in motion, they generate magnetic fields of force, so that electric and magnetic fields of force exist at the same time. Hence the expression *the electromagnetic force,* indicating that electricity and magnetism are intimately related and different facets of a single phenomenon. In about 1840 the great British physicist Michael Faraday (1791–1867) discovered that a moving magnet exerts a force on electric charges that causes them to move. He found that when a magnet is moved past a coil of wire, the electric charges in the wire move, so that a current is set up in the wire. This discovery, which is called electromagnetic induction and is the basis of the electric generator, completed what Oersted had begun: electricity plus motion gives magnetism, and magnetism plus motion gives electricity. From this simple dualism comes all modern electromagnetic technology.

From the discoveries of Oersted and Faraday the great Scottish mathematical physicist James Clerk Maxwell (1831–1879) deduced by pure mathematical reasoning that the electromagnetic field can be made to move out into space at the speed of light—186,000 miles per second—in the form of a wave consisting of a rapidly oscillating electric field and a rapidly oscillating magnetic field. Such a moving electromagnetic force field will cause any electric charges or magnets that it encounters to oscillate in unison with the original oscillating charge that set up the electromagnetic wave, just as a cork on water is set oscillating by a wave of water. If an electric charge is oscillated back and forth about 500 trillion times per second, the electromagnetic force field that rushes away from it and enters the human eye causes the electric charges on the retina to oscillate at the same rate and the brain tells one that he is seeing red light; if an electric charge is oscillated 1,000 trillion times per second, it generates an electromagnetic wave that is called blue light; and so on.

When a radio or television set is tuned, the electric circuitry in the set is arranged to oscillate in response to electromagnetic waves that are generated continuously by oscillating electric charges in the radio or television stations. Every time someone sees an object, the electric charges on the retinas of his eyes are responding to the electromagnetic force fields that are generated by the oscillating charges in the object. When he beholds a distant star, the electric charges in his eyes are oscillating in unison with an electromagnetic force field that originated from oscillating charges in the star many years ago and spent those years traveling across the vast space between the earth and the star.

Not only do electric charges in the eyes respond to electromagnetic force fields but so also do the electric charges in the skin. Charges in the eyes respond to such fields only if the fields are oscillating at a rate that is anywhere between 500 and 1,000 trillion times per second. When the skin feels the warmth of the sun, the charges in the skin are responding to infrared light, the electromagnetic force fields from the sun that are oscillating about 300 or 400 trillion times per second. When the skin tans, the charges in the skin are responding to ultraviolet light, electromagnetic force fields that are vibrating more than 1,000 trillion times per second.

The complexity of the electromagnetic force field gives rise to a wonderful variety of phenomena. All life processes—

thinking, digestion, muscular responses, growth, sexual responses—all chemistry, and all optical phenomena arise from this force.

Since, under ordinary circumstances, the objects surrounding man are electrically neutral, it is difficult to accept the idea that all bodies are composed of vast numbers of individual electrical charges, but that is the case. All matter consists of equal numbers of negative and positive charges, and that is why matter ordinarily is neutral. The basic unit of negative electric charge is the electron; it is a tiny particle of matter whose mass is so incredibly small that it would take about 1,000 trillion trillion (1 followed by 27 zeros) electrons to make up one gram of these basic particles. Of course, a gram of electrons could never be concentrated in a single small particle, because the force of repulsion between the electrons would rip the particle apart. The basic unit of positive electric charge is the proton; it, too, is a very tiny particle, but its mass is about 2,000 times that of the electron, so that protons, together with certain neutral particles called neutrons, account for the mass of matter. All atoms consist of equal numbers of electrons and protons, so that all electromagnetically neutral matter consists of equal numbers of protons and electrons. One gram of hydrogen contains about 600 billion trillion (6 followed by 23 zeros) protons and the same number of electrons. In spite of these large numbers, one of the amazing facts about electric charge is that there is an exact equality between the number of electrons and protons in the universe. If there were as much as one percent difference between the number of negative charges and positive charges inside one ounce of ordinary matter, this bit of matter would be torn asunder by a force equal to the total weight of the earth. One other important point is that electric charge can never be created or destroyed.

The protons and neutrons in an atom are concentrated in a tiny central region about one ten-trillionth of a centimeter across called the nucleus (1 inch equals about 2.54 centimeters); the electrons circle around the nucleus in orbits whose diameters are about 100,000 times larger than that of the nucleus. The circling electrons are held in their orbits by the electromagnetic forces exerted on them by the positively charged nucleus. Thus, the force that accounts for the structure of the atom is the electromagnetic force, which is enormously stronger than the gravitational force between the electron and the nucleus. In fact, the electric attraction between the nucleus and an electron is 10,000 trillion trillion trillion times as large as the gravitational attraction between them. Since the nucleus of an atom is much smaller than the atom itself, it is quite correct to conclude that nuclear forces are much stronger than the electromagnetic force.

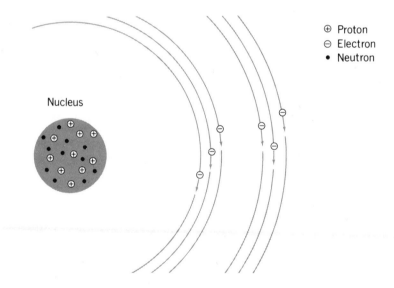

The electromagnetic force between the positively charged nucleus and the negatively charged electrons of an atom keeps the electrons revolving around the nucleus in their orbits. The short-range but powerful nuclear force keeps the protons and neutrons together in the nucleus.

NUCLEAR FORCE

The electromagnetic force keeps an atom together in the sense that it keeps the atomic electrons revolving around the nucleus of the atom, but it is the nuclear force that keeps the nucleus itself together. Although the nature of the nuclear force is still a good deal of a mystery, some of its properties were discovered in the early 1930s. It is now known how the nuclear force operates inside the nucleus and how it keeps the protons and the neutrons in the nucleus together. That a very strong force is required to keep the particles in the nucleus of an atom together is evident from the fact that about half of these particles are protons, which give the nucleus its positive electric charge and about half its mass. Since the protons in the nucleus are extremely close together, the nucleus would be blown apart by the mutual electric repulsion of these protons if there were not some other extremely strong attractive force holding them together. In fact, the early atomic bombs—uranium fission bombs—got their tremendous explosive power from the electric repulsion of the protons in uranium nuclei, which, under certain conditions, can overcome the nuclear attraction.

Since physicists knew that large electric repulsions must exist between the protons inside the nuclei of all atoms, they were greatly puzzled about the stability of these nuclei. From 1911 to 1932 one of the most challenging questions that faced physicists was, What keeps the nucleus of an atom from exploding? The answer came in 1932 with the discovery of the neutron, an electrically neutral particle, which resides in the nucleus of an atom and which is just slightly more massive than the proton. Experiments in which neutrons and protons were made to collide with each other and with nuclei demonstrated the existence of an extremely strong attractive force, the nuclear force, between all nuclear particles. Although the nuclear force is many times stronger than the electric force of repulsion between the protons in the nucleus, its range is extremely small in comparison to the ranges of the gravitational and electromagnetic forces. It is only when a proton or neutron (these particles are called nucleons) comes to within one-tenth of a trillionth of a centimeter of another neutron or proton that these particles are pulled to each other by the nuclear force. That is why the nuclei of atoms are so very tiny compared to the atom itself.

Under ordinary circumstances, the very short range of the nuclear force prevents the nucleus of one atom from exerting any nuclear force on the nucleus of another atom. Only when the nuclei of atoms get very close together, as in the deep, hot interiors of stars or in hydrogen bombs, do they attract each other strongly and coalesce to form heavier nuclei; this process is called thermonuclear fusion. Matter is ordinarily extremely stable because the electric repulsion between the nuclei of atoms is much too strong to permit them to get close enough for the nuclear force to come into play and bring about fusion. The stability is upset by free neutrons— that is, neutrons not bound in a nucleus—because neutrons have no electric charge and therefore are not repelled by the positive charge on atomic nuclei. A neutron can thus get very close to a nucleus and be dragged into it by the nuclear force. When this happens, the equilibrium in the nucleus is upset and the nucleus changes its character; it may become radioactive, emitting a charged particle, and thus change into another nucleus. When a free neutron enters a uranium or plutonium nucleus, fission of the nucleus occurs, which is the basis of the early fission atomic bombs. Fortunately for life on earth, free neutrons are quite rare because they are unstable; a free neutron left to itself decays into a proton, an electron, and another particle called a neutrino in about half an hour, so that an accumulation of neutrons is impossible. If this were not so, vast numbers of neutrons would surround man and would alter the chemical elements in his body by entering the nuclei of his atoms. The resultant radioactivity of these nuclei would then destroy his body cells. That is why a neutron bomb—an atomic bomb that emits large quantities of neutrons—is so dangerous to all living things. Although free neutrons are unstable, neutrons are quite stable when they are inside nuclei, where they contribute to the stability and mass of the nuclei themselves.

Neutrinos are mysterious, electrically neutral, spinning particles that travel at the speed of light and are present in great numbers everywhere in the universe. Because they are electrically neutral, they pass through the earth, the planets, and the stars with great ease. These mysterious particles are the only known bits of matter other than the photons of light (that is, light itself) that travel at the speed of light. Neutrinos have no intrinsic mass of their own but derive their mass and inertia from their energy; they have an infinite lifetime.

WEAK INTERACTION

The fourth and most mysterious of the four known forces of nature is the weak interaction, as it is called by physicists. The weak interaction force, which is operative between protons, electrons, and neutrinos, is responsible for those very rare situations in which a proton, an electron, and a neutrino come together to form a neutron. Since this force is very weak compared to the electromagnetic force but still very much stronger than the gravitational force, the chance that protons, electrons, and neutrinos will get together to form neutrons is very small. Little else can be said about weak interaction, for it is still one of the mysteries that physicists are trying to penetrate.

An understanding of the four known forces of nature is essential for an understanding of the structures in the universe and the manner in which these structures evolved in the course of 13 billion years from the primordial, undifferentiated matter. Each of these forces is dominant in a different domain; the nuclear force dominates the nucleus of an atom but plays hardly any role in the structure of the outer part of the atom or in the structure of the solar system; the electromagnetic force is dominant between the nucleus and electrons of atoms and between atoms in molecules but is relatively unimportant in the nucleus of atoms or in the motions of the planets; the gravitational force governs the motions of the stars but appears to be quite unimportant inside the atom or in the nucleus.

2 Galaxies and the Expanding Universe

Any discussion of the "origin" of the universe requires an understanding of what the word *origin* means to the cosmologist. To the nonscientist it means how the matter in the universe came into being, but to the cosmologist it means how the universe as it is now evolved from some initial state in which matter was already present and in which the total amount of matter was the same as it is now. The layman will often ask the astronomer where the matter in the universe came from, and the astronomer's answer, as it should be, is, "I do not know; that is not a question that really concerns me because there is no way scientists can answer that question." To the astronomer and all other scientists interested in this problem the story of the origin of the universe begins with matter already in existence, but in a form quite different from its present one.

The Milky Way

To observers of the heavens in ancient civilizations the universe appeared to be eternal and unchanging; the visible stars were arranged in seemingly fixed patterns, constellations whose configurations showed no observable alterations during the lifetimes of these observers or even during the lifetimes of some of these civilizations. People therefore quite naturally accepted the idea that the universe had been created by some divine force just as it appeared to them and that it would continue that way throughout eternity. The concept of a divinely created, unchanging universe dominated man's thinking for thousands of years, and as late as the first half of the nineteenth century so sophisticated a poet as Goethe expressed this in his introduction to his poetic drama *Faust*:

> The sun makes music as of old
> Amid the rival spheres of heaven,
> On its predestined circle rolled
> With thunder speed: The Angels even
> Draw strength from gazing on its glance,
> Though none its meaning fathom may;
> The World's unwithered countenance
> Is bright as at creation's day.

The idea of an evolving universe is a twentieth-century concept that grew out of the vast accumulation of observational data that became available only when the large, modern telescopes were constructed and focused on the stars. Without these telescopes astronomers could not have observed the distant galaxies and discovered that they are receding, which was the first important clue that led to the present picture of an expanding universe, in agreement with the predictions of Einstein's famous general theory of relativity.

Before the observational evidence that the galaxies are receding from earth and from each other is presented, the structure of a galaxy must be described, with the Milky Way, earth's galaxy, serving as an example. On any clear moonless night, the sky appears to be covered with stars, but the fact is that only about 3,000 individual stars are visible to the normal naked eye in either the northern or southern hemispheres. These stars are visible without the aid of a telescope because they are either very luminous or relatively close to us. *Relatively close* here means that the distances of most of these stars are small compared to the size of the entire collection of stars called the Galaxy or the Milky Way. The star closest to earth—excluding the sun—is Alpha Centauri, in the constellation of Centaurus; its distance from earth is such that the light coming from it, traveling at 186,000 miles per second, takes about 4.5 years to reach earth. Alpha Centauri is therefore said to be at a distance of 4.5 light-years, or about 26 trillion miles (1 light-year is about 6 trillion miles). All the stars that can be seen with the naked eye lie within a distance of about 1,000 light-years from earth. They constitute only a very small fraction of the stars in the Milky Way, most of which, because of their great distances from earth, are too faint to be seen individually; however, taken all together, they are visible as a broad luminous band in the night sky that looks like a thin cloud. The Milky Way circles the entire sky, half of it being visible in the southern hemisphere of the sky and the other half in the northern hemisphere.

It was long thought that the visible stars, which are distributed over the entire sky as viewed from the earth, are part of a structure quite distinct and different from the Milky Way, which, as seen from the earth, is concentrated in a relatively small band that cuts across the sky. In the eighteenth century it became clear through the observations of the great Sir William Herschel (1738–1822), a German musician turned British astronomer, that the Milky Way is composed of innumerable stars and that these Milky Way stars and the few thousand nearby visible stars form a single stellar system. Herschel surmised that this is indeed so and was the first to propose the idea that the Milky Way is a thin, disklike structure bulging at the center like a convex lens. Even a small telescope reveals many of the individual stars in the Milky Way to a viewer, and photographs of the Milky Way taken with the 100-inch telescope at Mt. Wilson Observatory, near Pasadena, California, or the 200-inch telescope at Mt. Palomar Observatory, also in California, are crowded with millions of star images, so that they merge into each other on many parts of the plate, forming what appear to be vast clouds of stars. Owing to this cloudlike appearance, astronomers refer to the piling up of stars within the Milky Way as star clouds.

With modern photographic techniques and special microscopes to count the star images on photographic plates, astronomers have verified that the visible stars and those in the Milky Way that can be seen only through a telescope or on photographic plates are part of a single stellar system that will hereinafter be called the Galaxy. The observational data show that there is a gradual thinning out of stars as one looks away from the plane of the Milky Way and a great increase in the concentration of stars toward the plane of the Milky Way. Moreover, the distribution of stars, as revealed by telescopic photographs, shows clearly that the visible stars that form the well-known constellations merge gradually into the star clouds of the Milky Way, so that they are part of the Milky Way system also.

Why then do the visible stars in constellations appear to be distributed fairly uniformly over the entire sky, whereas the Milky Way is concentrated in a band that cuts across only one part of the sky? There is a twofold reason for this. First, the sun is in the midst of the nearby stars, so that they appear to surround earth. Second, the sun is not at the center of the Galaxy, as Herschel had thought, but near the edge of the Milky Way or the Galaxy, so that man gets a one-sided view of the Milky Way. The solar system is about 30,000 light-years from the center of the Galaxy, with the result that when one looks toward the center, in the general direction of the constellation Sagittarius, the stars are so crowded together there that they appear to be concentrated in a band rather than spread out over the entire sky, as they would be if earth were near or at the center of the Galaxy.

The analysis of the numbers of star images on innumerable photographic plates, covering every part of the sky, has shown that the Galaxy is a vast spiral structure of more than 100 billion stars, consisting of a concentrated core of the very oldest stars and of three distinct spiral arms that originate in the core and contain the young and middle-aged stars. The core has a diameter of about 30,000 light-years and a thickness of about 15,000 light-years at its center, where the concentration of stars is estimated to be anywhere from 100,000 to 1 million times greater than in the spiral arms, where are found the younger and middle-aged stars, such as the stars in the constellation of Orion and the sun. Each spiral arm of the Galaxy is about 1,000 light-years thick (the dimension perpendicular to the galactic disk) and 3,000 or 4,000 light-years wide (the dimension in the plane of the Milky Way measured radially). The spaces between the spiral arms are also a few thousand light-years wide. Whereas the core of the Galaxy is thickly populated with stars, the spiral arms are quite thinly populated; if man lived in the core, there would never be any real nighttime for him, owing to all the starlight there. The distance between neighboring stars in the arms is about 4 or 5 light-years. The arms are also distinguished from the core by the presence of grains of dust between the stars in the arm—ice particles, large organic molecules, and the like—which obscure the center of the Galaxy from direct view. This dust and the hydrogen and helium gas in the arms are the raw material from which new stars are born.

The sun and its planets lie in the middle spiral arm of the Galaxy—called the Orion arm because the constellation Orion lies within it—near the inner edge of this arm. The spiral arm closest to the core of the Galaxy is called the Sagittarius arm because the constellation Sagittarius lies in it; the outermost spiral arm is called the Perseus arm because the constellation Perseus lies in it. The three spiral arms and the core together form a vast spiral structure of stars having a diameter of about 80,000 light-years.

The very oldest stars, stars at the core of the Galaxy, are about 8 or 9 billion years old. The recently born stars, in the spiral arms, are at most a few million years old and have been dubbed the "kittens of the Milky Way." The vast core of the Galaxy is characterized by one other interesting feature: it is surrounded by a halo of globular clusters. Over a hundred of these remarkable structures, each almost perfectly spherical and each containing anywhere from 100,000 to 1 million stars, have been discovered lying in a shell surrounding the nucleus of the Galaxy. Like the stars in the core of the Galaxy, the stars in the globular clusters are very old.

Motions of the Stars

To the ancients, and even to men like the Polish astronomer Nicolaus Copernicus (1473–1543), the Danish astronomer Tycho Brahe (1546–1601), and the German astronomer Johannes Kepler (1571–1630), who ushered in the modern era of the study of the heavens, the stars appeared to be fixed in the sky, maintaining their relative positions for eons. In fact, the expression *the fixed stars,* which is still commonly used even though it is now known that all the stars in the Galaxy are moving about, is a carry-over from ancient times. That the stars must be moving about in some fashion was already clear to Newton when he discovered the law of gravity. According to the law of gravity, every particle of matter in the universe pulls on every other particle. All the stars in our Milky Way pull upon each other, but owing to the vast distances between even neighboring stars like the sun and Alpha Centauri, the pull of any single star on any other is extremely small and hardly detectable. In spite of this, the stars in the Galaxy do influence each other noticeably because of the vast numbers of stars that are involved. If one keeps in mind that the sun, like all the other Milky Way stars, is subject to the gravitational pull of billions of stars, one can accept the idea, as did Newton, that the sun and the other stars must be moving under the action of the force of gravity. Since stars are free to move about, they must respond to the gravitational pull of all the other stars and move about in such a way as to keep the Galaxy stable and give it its observed structure, just as the planets move in their orbits around the sun under the gravitational action of the sun and thus give the solar system its observed structure.

With the aid of large telescopes after the time of Galileo, Herschel and other astronomers detected and measured the motions of individual stars by noting that over periods of years the stars shifted their positions relative to the other stars in the same parts of the sky. Careful measurement of the positions of the images of the few thousand nearby stars on photographic plates year after year showed that the stars in the neighborhood of the sun are moving about, relative to the sun, at speeds of 20–30 kilometers (13–20 miles) per second. Although these motions at first appeared to be random, ingenious analyses by twentieth-century astronomers showed that a definite pattern in these motions confirms the picture of the Milky Way as a vast spiral structure.

The motions of the stars were at first measured relative to the sun, as though the sun were fixed in the Galaxy. But this, clearly, cannot be the case; the sun must be moving the way the stars in its neighborhood are moving. In fact, astronomers discovered that, relative to the stars in its own neighborhood, the sun is moving—and carrying the planets with it—at about 13 miles per second toward a star in the constellation of Hercules. With the discovery and the analysis of the sun's apparent motion relative to the nearby stars, astronomers, within the last fifty years, have been able to find the overall pattern of the stellar motions and to relate this to the structure of the Galaxy. The stars, including the sun, are all moving in vast elliptical orbits around the core of the Galaxy in just the way the planets are moving around the sun and for the same reason: gravity. If this were not so, all the stars in the Milky Way would have coalesced into the core of the Galaxy billions of years ago to form a single spherical object instead of the flattened spiral structure that exists. Analysis of the observed stellar motions shows that the solar system and the stars in its

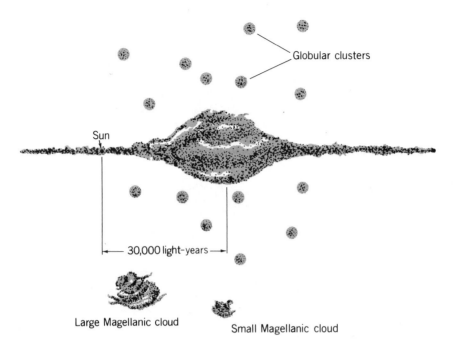

The Milky Way seen edgewise, showing the halo of globular clusters and the dark band of obscuring matter across the center. The Magellanic star clouds and the position of the sun are also shown. The sun is 30,000 light-years away from the center.

neighborhood are revolving around the core of the Galaxy once every 250 million years. Stars that are closer to the core than the sun is revolve more rapidly than the sun, and those farther away from the core revolve more slowly.

The revolution of the stars around the galactic core is governed by the force of gravity and accounts for the shape of the Galaxy and its spiral arms. Since the stars are kept moving in their orbits around the core of the Galaxy by the gravitational pull of the core, a simple formula derived from Newton's law of gravity can be used to calculate the total mass of the core. It can be shown that the speed of a star around the core depends on the square root of the total mass of the core, divided by the distance of the star from the center

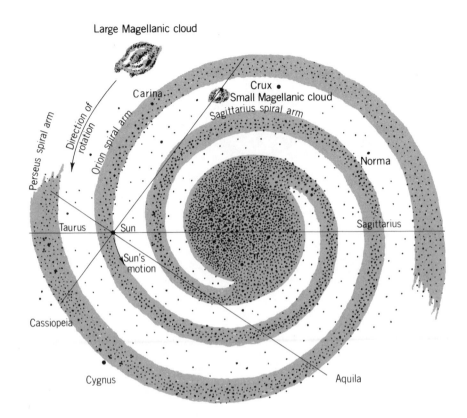

The position of the sun in the Milky Way, seen head on. The Galaxy rotates, as shown, once every 250 million years. As the result of this rotation the solar system revolves around the center of the Galaxy once every 250 million years, moving at a speed of 220 kilometers per second.

of the core. Since the speed of the sun in its orbit around the core and its distance from the center are known, this formula can be applied to calculate the mass of the core, which turns out to be about 100 billion times greater than that of the sun. It may thus be concluded that the core contains about 100 billion stars, which agrees with the estimate of the total number of stars in the Galaxy obtained from star counts.

The collective motions of the stars are described as the

rotation of the Galaxy. The Dutch astronomer Jan Hendrik Oort in 1927 deduced, by a careful analysis of the way the stars in different parts of the sky are receding from or approaching the sun, that the Galaxy is rotating. This was a most remarkable deduction, for it was made long before the structure of the Galaxy had been correctly analyzed and before the distance of the sun from the center was known. In recent years astronomers have completely confirmed Oort's conclusion by using radio telescopes to pick up the radio signals emitted by the electrons in the free hydrogen atoms in interstellar space. Since these hydrogen atoms lie in the spiral arms of the Galaxy, the hydrogen radio signals picked up by radio telescopes permit the tracing-out of the spiral arms and the determination of how the material in these arms is moving. These radio data completely confirm the spiral structure of the Galaxy and its rotation as deduced from optical data.

Radio telescopes give a much more detailed picture of the Galaxy than optical telescopes can. The reason is that the dust in the arms of the Galaxy obscures the center of the Galaxy and most of the core from direct viewing with optical telescopes. In fact, individual stars cannot be detected optically at distances greater than 6,000 light-years in the direction of the center of the Galaxy because of the interstellar dust. It is like trying to see the individual lights in a heavily fogbound city at night. But radio signals from the center of the Galaxy can be detected with radio telescopes, because the signals are no more affected by the dust than by a heavy fog. Light and radio signals are both electromagnetic waves; but the lengths of light waves are extremely short, whereas radio waves are long. Hence, tiny particles of dust or fog prevent the short light waves from getting through but have very little effect on the long radio waves, which flow around tiny particles. Radio signals from the center of the Galaxy have revealed a great deal of activity there that is still not fully understood.

Recession of the Galaxies

The relationship between the Milky Way and the external galaxies—or spiral nebulas, as they were first called—was not understood in the early years of modern astronomy because it was not clear that the external galaxies are outside the Milky Way and not part of it. Such spiral nebulas as the great nebula in the constellation Andromeda had been observed for almost 200 years before enough observational evidence was obtained to prove conclusively that they are at vast distances from earth compared to the distances of stars in the Galaxy. Whereas the nearest star, Alpha Centauri, is about 5 light-years from earth, the nearest galaxy is about 1 million light-years away, and the spiral nebula in Andromeda about 2 million light-years. There are some 30 galaxies, each containing billions of stars, within a radius of about 2 million light-years of the Milky Way, and these galaxies form a cluster of galaxies called the local cluster. The 30 or so galaxies in the local cluster are held together in this cluster by the force of gravity, just as the individual stars in a galaxy are held together. The force of gravity thus operates to form galaxies from stars and clusters of galaxies from individual galaxies.

From the data obtained with the large optical telescopes that were built all over the world during the last half century, astronomers have been able to analyze spiral nebulas in detail and to show that they have the same structure and composition as the Milky Way. In fact, the detailed study of the Milky Way has given us a deeper understanding of spiral nebulas, and the study of spiral nebulas in turn has confirmed the picture of the Milky Way as a spiral structure. Photographs of many spiral nebulas, which present themselves in various orientations, show all the structural features of the Milky Way. A single example, the spiral nebula in Andromeda, illustrates this point nicely (see picture section). This famous

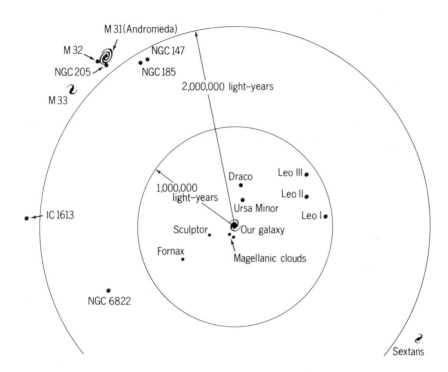

The local group of galaxies. Only three galaxies in the group are spirals: the Milky Way, Andromeda, and M33.

galaxy, which is so oriented that it is seen almost head-on, the core and spiral arms being easily visible, is about twice as large as the Milky Way and has about twice as many stars. The spiral arms (about five of them) stand out quite clearly, and dust lanes can be traced out along these arms. The stars in these spiral arms are much younger than, and appear quite different from, those in the core: the stars in the arms are bluish, whereas those in the core are reddish yellow. The entire galaxy is rotating very slowly (about once every few hundred million years), with the individual stars in the spiral arms revolving around the core in huge elliptical orbits.

The real nature and structure of the spiral nebulas could not be determined until their distances were known. This became possible when it was discovered that they contain certain pulsating stars, known as Cepheid variables, which have the property that the more slowly they pulsate, the more luminous they are. Knowing the luminosities of the Cepheid variables from their rates of pulsation, astronomers can determine the distance of the variables by comparing the known luminosities with the apparent brightness of the variables. In this way astronomers, in the third decade of this century, discovered that most of the known spiral nebulas are millions of light-years from earth. Although this fact had already been surmised by most astronomers before the distances had actually been determined, the measurement of these distances revealed to man that he could now determine the structure and composition of the universe itself. With this exciting prospect in mind, astronomers in even greater numbers began to study the spiral nebulas in more and more detail.

Increasing study soon revealed two things. First, in addition to existing in the form of spiral structures, external galaxies exist in the form of huge ellipsoidal and spherical structures with no spiral arms, and even as irregular structures. Second, the external galaxies tend to arrange themselves in groups or clusters that are held together by the force of gravity. A remarkable example of this clustering is the famous Virgo cluster of galaxies, which is some 40 million light-years away and which consists of some 2,500 galaxies of all types, ranging from the spherical to the irregular.

Just as the motions of the stars within a given galaxy provide deeper insights into the dynamics and structure of that galaxy than could the distribution of the stars alone, so the motions of the galaxies and clusters of galaxies provide deeper insights into the dynamics and structure of the universe than can the distribution of the galaxies alone. Since the galaxies are very far away, their motions could not be detected by the eighteenth-century astronomers in the way the motions of the nearby stars could. While careful observations of the neighboring stars year after year by Herschel and others showed that the stars change their positions relative to each other and hence must be moving, there is no observable change in the positions of the galaxies, which means either that the galaxies are motionless or that they are so far away that their motions, even though rapid, give rise to no perceptible change in their positions even after many years of observation. A third possibility is that the galaxies are all moving radially, along the line of sight, so that no shift in positions occurs.

As long as astronomers could only determine the motion of celestial objects by measuring their lateral or transverse displacements in the sky—that is, displacements at right angles to the line of sight—there was no way for them to discover how, or whether, the galaxies are moving. But then an important property of light that permits one to measure radial motions was discovered by the Austrian physicist Christian Doppler (1803–1853), the famous Doppler effect. Since white light is a mixture of pure colors ranging from red to violet and since light of all colors consists of electromagnetic waves that are propagated in empty space at a constant speed of 186,000 miles per second, all colors qualitatively represent the same kind of physical phenomenon: the propagation of electromagnetic vibrations. Two different electromagnetic vibrations (that is, two different rays of light) differ in color if the wavelengths of the vibrations (that is, the distances between two successive crests of the rays) are different. The longer the wavelength, the redder the light. The wavelengths of the visible colors are very short, ranging from about 3.5/100,000 of a centimeter for violet light to twice that value for red light. Longer wavelengths than this second value correspond to infrared rays, heat rays, and radio waves; wavelengths shorter than the first value correspond to ultraviolet and X rays.

Doppler discovered the remarkable fact that the wavelength, and hence the color, of the light entering a person's eye depends on the motion of the observer relative to the source. If a source of light is receding—that is, if the distance between source and observer is increasing—the light appears redder than it would be if the distance between source and observer were constant. On the other hand, if the source is approaching the observer, the wavelength of the light appears bluer than it would be if the distance between source and observer were fixed. In other words, if a source of light is moving radially relative to an observer, the wavelength of the light is altered. Since the amount by which the wavelength of light increases or decreases depends, respectively, on the speed of recession or approach of the source, one can determine the speed of recession or approach of a source of light—that is, its radial velocity—by measuring the change in the wavelengths of the various kinds of light emitted by the source. For this purpose, one must know what the wavelengths of the various kinds of emitted light would be if the source were not moving. Fortunately, this is known because the source consists of atoms whose radiation characteristics have been discovered in laboratory experiments performed with atoms on the earth. The radiation characteristics of an atom are called the spectrum of the atom. Since the spectra of all atoms on the earth—and these are the same atoms as are found on and inside stars—have been studied in detail, scientists know what the wavelengths of the radiation from these atoms are when the atoms are not approaching or receding. If one finds that the spectral patterns of atoms from a source of light like a galaxy differ slightly from the known patterns because all the wavelengths in the patterns are longer (or shorter) than they are for atoms fixed here on earth, one knows that the galaxy is receding from (or approaching) earth.

In 1912 the American astronomer Vesto Melvin Slipher used the Doppler effect to measure the velocity of the spiral nebula in Andromeda and found that it is approaching earth at a speed of 130 kilometers or 80 miles per second. A few years later Slipher analyzed the spectra of about 20 other

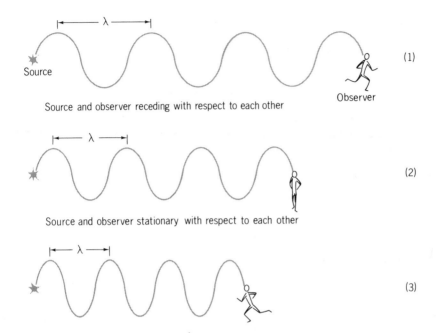

The change in the wavelength of light (the Doppler effect) as measured by an observer receding from the source of light and one approaching the source of light, compared with a stationary observer.

  1) Receding observer: wave length increases
  2) Fixed observer: wave length unchanged
  3) Approaching observer: wave length decreases

spirals and discovered two surprising things. First, these galaxies all show large Doppler shifts, and except for 2 galaxies, the Doppler shifts are toward the red, so that they are all receding from earth. Second, the fainter and smaller (in appearance) galaxies, and hence the more distant galaxies, have the largest Doppler shifts and hence the largest speeds of recession. Slipher found that the radial velocities of these distant galaxies are 20 to 30 times larger than those of the stars in the Milky Way and in some cases as large as 1,800 kilometers or 1,135 miles per second. In 1924 Slipher published his analysis of the spectra of 43 other galaxies, which completely confirmed his previous conclusions: the distant galaxies are all receding from earth, and the speed of recession increases with distance.

Since the distances of the faint galaxies were not known in 1924, nothing could be said at that time about the relationship between the speeds of recession of these galaxies and their distances, but by the 1930s this was possible as the result of the work of two American astronomers, Milton L. Humason, a janitor and handyman turned astronomer, and Edwin Powell Hubble (1889–1953). The unexpected and unexplained sizes of the radial velocities of the distant galaxies, measured by Slipher, led Humason in 1929 to undertake with the new 100-inch Mt. Wilson telescope a systematic investigation of the spectra of all the known distant galaxies. So faint are some of these very distant galaxies that Humason had to use exposures of 50–100 hours on some of his photographic plates to obtain the images he needed for his spectral analysis of the light and the measurement of the Doppler effect. His results confirmed the work of Slipher. The fainter and smaller the images were, and hence the more distant the galaxies were, the greater were the attendant Doppler shifts and hence the speeds of recession. By 1935 Humason had found a galaxy in the Ursa Major cluster of galaxies with a Doppler shift large enough to correspond to a speed of recession of 42,000 kilometers per second. In recent years far larger Doppler shifts have been measured.

Humason's work stimulated Hubble, who devoted his life to galaxies, to measure the distances of the galaxies whose spectra Humason had analyzed. Using a variety of ingenious techniques, Hubble discovered that the speeds of recession of the galaxies are strictly proportional to their distances from earth. This is known as Hubble's law of recession; it tells us that if galaxy A is twice as far from earth as galaxy B, its speed of recession is twice that of B. This law of recession can be stated numerically by introducing Hubble's constant, a number that signifies how the speed of recession of the galaxies increases with distance. Hubble's data on the distances of the galaxies gave a large value for this constant, but since then, particularly within the last 15 years, with improved techniques for measuring these distances and with the use of the 200-inch Mt. Palomar telescope, astronomers have deduced a value of about 24 kilometers per second per million light-years for the Hubble constant. This means that the speed of recession increases by 24 kilometers per second for each million light-years' increase in distance. Thus, a galaxy or cluster of galaxies at a distance of 10 million light-years is receding at a speed of 240 kilometers per second, whereas galaxies 100 million light-years away are receding at 2,400 kilometers per second, and so on.

With the use of the 200-inch telescope and with improved photographic and spectroscopic techniques of recent years, astronomers have measured the Doppler shifts of the galaxies out to distances of a few billion light-years and have found Hubble's law still valid. However, one other important discovery was made: the value of the Hubble constant today is smaller than it was in the past. The spectroscopic evidence shows that speeds of recession of the galaxies today are smaller than they were in the past; in fact, there has been a steady drop in the value of Hubble's constant, assuming that the spectroscopic evidence has been correctly interpreted. This point has an extremely important bearing on the understanding of the origin and evolution of the universe.

The General Theory of Relativity

The discovery that the spectra of the distant galaxies are all red-shifted and that these red shifts increase with increasing distances of the galaxies was greeted with mixed reactions and with various interpretations by scientists and nonscientists alike. Some scientists were quite surprised at the discovery of the red shifts and were very reluctant to accept the Doppler

interpretation, even though this discovery was not unexpected, for it had already been predicted by the work of
Einstein in 1917. But the deductions stemming from Einstein's
work were so amazing and revolutionary that many scientists
dismissed them as the sheerest speculation, protesting that
they did not correctly describe the universe. But, as will be
shown, Einstein was correct, and his ideas prevail.

Here one must consider the general question of the
validity of theoretical deductions made from a theory that is
correct. Owing to the vast changes that have occurred in
technology during the last half century and the many
controversies among scientists that have flowed from the
discoveries of new facts about the universe, there is a general
feeling among nonscientists that scientific theories are not to
be trusted, are the least stable of all intellectual creations, and
change almost as rapidly as the weather. This is not so at all.
This misconception about the endurance of a theory in
science stems from a misunderstanding of the nature of a
scientific theory. Most people confuse the interpretation of a
set of facts with a theory, and that is why the general mistrust
of theories and the misconception of the life of a theory have
arisen. Theories are the most stable of all the constructs of the
human mind; they are clung to most tenaciously and rarely
die. Moreover, in spite of the fact that most people pride
themselves on being very practical, hard-headed, and distrust-
ful of "theories," there is hardly a thing they can do that is not
based on theory. The very interpretation and understanding
of a "fact" are influenced, and often determined, by theory.

A good example of the difference between a theory and
an interpretation of a fact is presented by the high temperature of the surface of Venus. When astronomers discovered
from the radio waves emitted by Venus that its surface is very
hot (about 800°F), they proposed various explanations. The
high temperature on Venus may be caused by the great
concentration of uranium on its surface (radioactivity); by the
entrapment of solar radiation in its atmosphere (the green-

house effect); or by the escape of heat from a hot, molten interior to the surface. Each of these statements is a possible explanation of why Venus is hot, but none is a theory in the sense that this term is understood by scientists. A theory in physics is a body of general laws that apply to all possible phenomena and not merely a statement about one particular phenomenon. Thus, one may speak of Newtonian theory, the theory of relativity, or the quantum theory, each of which encompasses a body of laws that must apply to every domain of the universe if the theory is correct. The various explanations of the high surface temperature of Venus proposed by scientists during the last ten years cannot all be correct, but they are all consistent with and, indeed, deducible from, the theory of relativity and quantum theory, which are the two great modern physical theories that have replaced Newtonian theory.

Newtonian theory has served mankind extremely well for hundreds of years in mankind's search for an understanding of the universe, and it is still very useful for many purposes. Newtonian theory is still accurate enough for the calculation of the orbits of the astronauts when they are sent on a trip through space; but when faced with questions involving bodies moving at high speeds or in intense gravitational fields, Newtonian theory breaks down and must be replaced by the theory of relativity. Again, questions dealing with electrons moving around inside atoms and molecules require that the Newtonian laws of motion be replaced by laws of motion derived from, and consistent with, quantum theory. Both the theory of relativity and quantum theory are twentieth-century theories that were born at the end of the nineteenth century and only gradually replaced Newtonian concepts of space, time, mass, and causality, which shows how enduring scientific theories are.

Returning to the consideration of the truth of deductions made from correct theories, consider an example from Newtonian theory that Newton himself carried out and that shows the great power of mathematical analysis when combined with a correct scientific theory. By the time Newton was born in 1642, Kepler, the "lawgiver of the solar system," had already discovered and stated the three laws of planetary motion, but these were empirical discoveries, based on some 30 years of trial and error and on very long and tedious numerical calculations. To achieve this, Kepler had to use the planetary observational data that Tycho Brahe had collected previously over a period of many years, so that Kepler's discoveries required the combined work of two men, extending over more than half a century. When Kepler stated his three laws of planetary motion, he did not know why the planets move the way they do but surmised that they do so because the sun exerts some kind of influence on them. Compare this now with Newton's great achievement. First, Newton stated his law of gravity, which asserts that the sun attracts each planet according to a very definite and precise mathematical formula involving the masses of the sun and the planet and the distance between them. From this formula alone, using no more than high school mathematics, he then deduced—in about an hour—all that Kepler had discovered in all his years of hard work. Moreover, the three laws of planetary motion, as deduced by Newton from his law of gravity, revealed a small, but important, error in Kepler's third law of planetary motion. This shows that theoretical derivations from a correct law are more trustworthy than empirical deductions. The reason for this is that a theoretical derivation is based on precise mathematical relationships and operations, so that if the basic law is correct, all mathematical consequences that flow from it must also be correct. Empirical derivations, on the other hand, are based on observations, which are bound to have errors—however small—in them, and these errors can lead to wrong conclusions.

Another important point in connection with this is that a theoretical derivation is more general than an empirical one, since one can never be sure that the empirical results apply to any phenomena not covered by the data involved. Thus, until Newton did his work, one could not be sure that Kepler's laws were valid for any bodies in the heavens other than the planets. But since Newton's law of gravity applies to all bodies, all deductions that stem from it apply to all bodies attracting each other gravitationally and not only to the planets revolving around the sun. Whether it be the moon or an artificial satellite revolving around the earth, a pair of binary stars revolving around a common center of mass, or a star revolving around the core of a galaxy, the deductions from Newton's law of gravity apply.

The emphasis here on the importance and permanence of theories in modern science is necessary because much of the discussion is based on the two great modern theories of physics: the theory of relativity and the quantum theory. If these theories are accepted as correct—which they must be at the present time, since no experimental or observational evidence has yet been presented that conflicts with them— then all predictions that can be deduced from these theories mathematically, however bizarre and contrary to common sense they may appear, must be accepted. The theory of relativity—both its special part, which Einstein published in 1905, and its general part, which he first published in 1915—is famous for such bizarre deductions. Thus, as an example, the special theory of relativity predicted the existence of antimatter—now completely confirmed—a number of years before particles of antimatter were discovered in the cosmic rays incident on the earth.

Among the predictions about the universe that stem from the general theory of relativity is one that lends substance to the interpretation that the red shifts of the distant galaxies are, indeed, Doppler effects and therefore tell us that the galaxies are receding from us and from each other.

After Einstein had formulated the general theory of relativity, which led to a new law of gravity and which introduced the concept that the space-time of the universe is a curved four-dimensional geometrical manifold, governed by the laws of non-euclidean geometry, he applied the theory to the structure of the universe in a famous paper in 1917. In this paper, which ushered in modern relativistic cosmology, Einstein showed that the curvature of space-time leads to a universe in which space curves back upon itself to form a closed universe that is finite but boundless. Einstein's initial model of the universe was a static, nonexpanding model; but shortly after Einstein published this remarkable paper, the Dutch astronomer Willem de Sitter (1872–1934) showed that Einstein's model of the universe is not stable; at the slightest disturbance it becomes an expanding universe, in which the galaxies rush away from each other at speeds that increase with distance. In fact, this deduction prompted astronomers like Humason and Hubble to investigate the distant galaxies, to see whether they are, indeed, rushing away from each other. The spectral data from the distant galaxies all show red shifts that increase with increasing distances of the galaxies. These observed red shifts agree with the predictions from Einstein's cosmology if one accepts the red shifts as Doppler shifts—that is, if one agrees that the red shifts mean that the galaxies are receding from us.

Although this is the most reasonable conclusion to draw from the data—indeed, there is no other known explanation —it was not immediately accepted by scientists. Even now, there is a small group of cosmologists who are doubtful about the recession of the galaxies and who seek explanations other than the Doppler effect for the red shifts. Although light can be reddened in ways other than the Doppler effect, the conditions required for such reddening do not occur in intergalactic space, so one is left with the Doppler effect unless, as some people have proposed, one posits some mysterious phenomenon in space that reddens light as it travels over long distances. But since the Doppler effect is a perfectly valid explanation for the red shifts, there is no need to appeal to mysterious forces that explain nothing and only confuse the issue.

Expanding Universes

However much one may contest the idea of receding galaxies and seek other explanations for the red shifts, one must, in the light of all evidence, accept recession as real and use it to construct an acceptable model of the universe. Although the general theory of relativity may lead to a closed and expanding model of the universe, it can yield expanding models that are not finite. In fact, in 1922, the Russian mathematician Alexander Alexandrovich Friedmann (1888–1925) showed that the model of the universe that Einstein first derived from his equations is but one of a series of possible models, all of which consist of receding galaxies but are not closed. Once one accepts the idea of an expanding universe and applies to such a universe Einstein's general relativistic equations, one obtains an array of possible theoretical models that agree with the overall observations of the recession of the galaxies but differ in certain important details. Since each of these models gives a different picture of the detailed way in which the universe originated, evolved, and will continue to evolve, one must determine which model is correct before one can give a definite and reliable picture of the birth of the universe and what will happen to it. This was virtually impossible just a few years ago because certain crucial observational data were missing; but now great technological developments in radio astronomy have made accessible the data necessary to choose the correct model, or at least one that best fits the observed data.

The expansion of the universe can be deduced from fairly general arguments if Einstein's "cosmological principle" is accepted: it states that the universe in the large looks the same for all observers, no matter where they are. If this is so, there is no region in the universe from which all the galaxies are receding or toward which all galaxies are rushing. If there were such a region, the concentration of galaxies there would be considerably different from their concentrations in other regions, and this would contradict the cosmological principle. If then, the galaxies are all receding from earth, as indicated by the red shifts, accepted as Doppler shifts, they must be receding from every point in the universe and hence from each other. In other words, the distance between the galaxies or between clusters of galaxies must be increasing. But this is possible only if the entire universe is expanding.

Here an analogy with the earth will aid in understanding this point. Suppose that all the people on the earth were distributed uniformly over the surface of the earth, including the oceans, which, for purposes of argument, are imagined to be solid, and that the people all around the observer are moving away from him. If this were true for all observers (the cosmological principle)—that is, if all observers on the earth saw people moving away from them—one would be forced to conclude that the earth is growing larger, like a balloon that is being blown up. If this were not so, the people running away

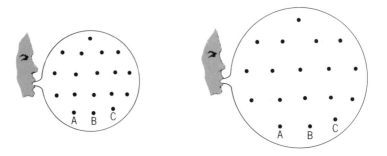

The surface of a balloon that is being blown up is a good model for the expanding universe. The individual galaxies (or clusters of galaxies) are to be pictured as disks on the surface of the balloon which do not change their sizes; only the distances between them (as measured along the surface of the balloon) change. Each disk may be pictured as the center of the expansion.

in all directions would be getting closer to observers on the other side of the earth. But if the earth were expanding radially, all distances along the surface of the earth would increase and all objects would appear to be separating from each other just like the distant galaxies. If, then, the three-dimensional universe is pictured as the skin of a four-dimensional balloon that is expanding along the fourth dimension, at right angles to the skin, and the galaxies as grains of sand stuck to the surface of the balloon, the galaxies will rush away from each other as the universe expands just as the grains of sand on the balloon recede from each other as the balloon expands. And so, if one accepts a finite, closed universe and the cosmological principle, then the red shifts must mean that the universe is expanding and the galaxies are receding from each other.

Not only are there finite expanding models of the universe but there are also expanding models that are infinite. Since the infinite expanding models go on expanding forever, whereas the finite expanding models stop expanding after a definite time and begin to contract, it is important to determine which of these models describes the universe.

3 Models of the Universe

Some Major Questions

Until quite recently only optical data such as the red shifts and the counts of the distant galaxies and clusters of galaxies (the number per unit volume of space) were available, but now it is possible to detect and measure, with telescopes designed just for these purposes, radio waves, infrared radiation, ultraviolet radiation, and X rays. In particular, the enormous growth of radio astronomy since World War II has revealed many features of the distant parts of the universe that could never have been discovered optically. The reason for this is that radio telescopes—huge parabolic dishes or arrays of dipoles—are much more sensitive than optical telescopes and can pick up incredibly weak radio signals. Moreover, whereas the sun is a strong emitter of optical radiation and thus blocks out the daytime stars, it is a weak radio emitter, compared to other celestial objects that are cosmologically important. Thus, earth receives about a thousand times as much radio energy every second from the Milky Way as it does from the sun and 100 million times as much visible radiation from the sun as from all other celestial bodies combined. Radio telescopes are so sensitive that they can detect a radio signal so weak that it would have to keep impinging on one square centimeter of the surface of the telescope for about 66 thousand trillion years to transport to that square centimeter a single calorie of energy (one calorie is the energy required to raise the temperature of one gram of water one degree centigrade).

There are four major questions whose answers are sought in the observational evidence:

(1) Is the universe an open, infinite system of galaxies that will continue expanding forever or is it a closed finite system that will stop expanding?

(2) Is the universe an unchanging universe; that is, has it always been the same and will it remain the same for all time, or is it an evolving universe whose characteristics are constantly changing and are quite different from what they were billions of years ago? This question became quite urgent some fifteen years ago when the steady-state model of the universe, with its necessary conclusion that matter is created continuously, was seriously proposed as the correct model.

(3) If the universe is an evolving universe—that is, if the steady-state model does not apply—when did it begin evolving and what was its initial state? Did it start out on its present life history gradually or was it launched into it violently?

(4) Finally, if the universe stops expanding at some time in the future, will it then contract again down to its initial, highly condensed state in a finite time and start another expansion phase or will it just go on contracting?

The Observational Evidence

To see what evidence must be presented to answer the first question, note that the force of gravity, acting continuously among all the receding galaxies, keeps slowing down the expansion as time goes on. If this deceleration of the expansion is strong enough, the expansion will be brought to a halt and the universe will start collapsing; otherwise, it will go on expanding forever. To see what kind of evidence must be collected in order to choose between these two possibilities, consider the simple problem of an object projected into space from the earth. If this object is shot into space at a speed greater than what is called the speed of escape (the speed needed to escape from the earth), it will never return to the earth, even though its speed decreases constantly as it moves away from the earth; otherwise, it will rise to a certain height and then fall back to the earth. The speed of escape from the earth is determined by the mass of the earth; the more massive a body is, the greater is the speed an object must start out with to escape from it. One can determine whether a body that is receding from the earth will escape from it, either by noting how rapidly its speed is decreasing from moment to moment, or by measuring the speed of recession of the body and comparing it with the speed of escape as deduced from a formula that is derived from Newton's law of gravity. Since this formula contains the mass of the earth, knowing the mass of the earth permits one to determine whether or not the body will escape.

A similar situation applies to the universe as a whole. Are the galaxies or clusters of galaxies rushing away from each other with speeds that are larger than the speed of escape for the universe as a whole or not? If they are, the universe is open and infinite and will go on expanding forever; if not, the universe is closed and finite and will stop expanding. If the universe began its expansion explosively at some moment in the past, so that the material that now constitutes the galaxies began moving off at high speeds in all directions at that moment, the present speeds of recession of the galaxies must be considerably smaller than the original high speeds of expansion, because of the steady slowing down of the expansion by the gravitational attraction of all the galaxies on each other. Since this slowing down of the expansion has been going on continuously from the moment the universe began its expansion, a comparison of the present speeds of recession of the galaxies with those of the past should show this and enable one to see whether the expansion will ultimately be halted. Such comparisons, which have been made in recent years by the American astronomer Allan R. Sandage, among others, are expressed in terms of a measurable quantity called the deceleration parameter. This number is simply the rate at which the Hubble constant is decreasing. Together with the Hubble constant, the deceleration parameter determines the dynamical properties of the universe.

To determine the value of the deceleration parameter, one must first compare the speeds of recession of galaxies that lie within a distance of about 100 million light-years with the speeds of recession of galaxies that are more than 1 billion light-years away. This will reveal how rapidly the speeds of recession have decreased and are still decreasing. This is possible because the light from the distant galaxies takes a finite time to reach earth, so that it does not reveal the way the galaxies look now or are receding now but rather the way they looked and were receding at various times in the past. In other words, the rays of light from the galaxies at different distances present pictures of the universe at different times in the past. The red shifts in the light from the galaxies within 100 million light-years give the value of the Hubble constant at the present time, or very nearly so, since not much of a change in the dynamics of the universe occurs in 100 million years; the red shifts in the light from the galaxies 1 billion or more light-years away give the value of the Hubble constant when the universe was 1 billion or more years younger. One can thus compare the present rate of expansion of the universe with its rate of expansion billions of years ago. This has been done, and the evidence shows that the galaxies are receding at speeds less than the speed of escape. The expansion of the universe must ultimately stop, which means that the universe is closed and finite; it is a four-dimensional space-time bubble that will never burst.

The Missing Mass and the Expanding Universe

Another way of determining how the universe is expanding can be deduced from a formula given by the general theory of relativity. This is an algebraic formula that expresses the Hubble constant in terms of the total mass in the universe and its present radius. It is fairly easy to understand the physical significance of this formula, because it is closely related to the formula for the speed of escape of a body from a spherical mass like the earth. Using Newtonian gravitational theory, one can show that if one divides twice the mass of the earth by its radius and multiplies the result by a certain physical constant known as Newton's universal constant of gravitation, one obtains the square of the speed of escape from the surface of the earth. This is written algebraically as follows:

$$V^2 = \frac{2M}{R} G,$$

where V is the speed of escape from the earth, M is the mass of the earth, R is the radius of the earth, and G is the famous Newtonian gravitational constant (6.7 divided by 100 million, in centimeter-gram-second units). Since this constant in Newton's law of gravity is extremely small, gravitational forces are in general very weak.

Suppose now that the speed of an object moving away from the earth is measured and found to exceed the value given by the formula—almost 7 miles per second—when the measured radius R and mass M of the earth are substituted into the formula. One would then expect the object to escape from the earth and not to return. But what if the object, instead of escaping as expected, returned to the earth? How could such a situation be accounted for? One would have to question the measured value of the mass of the earth and conclude that the earth is more massive than is apparent and that some of the earth's mass had somehow been overlooked. By underestimating the mass of the earth, one would arrive at a value for the speed of escape that is too small and then be puzzled by why objects moving faster than this value do not escape. One can apply the same ideas to the universe as a whole and see why the observed value of the deceleration parameter leads to a puzzle.

One can speak of the total mass M and the radius R of the universe. But one must be a bit cautious in interpreting R, because the phrase "the radius of the universe" as applied to R is meaningful only when the universe is closed and three-dimensional space can be regarded as the surface of a sphere in a four-dimensional euclidean space. If the universe is not closed, R is not really a radius but a changing cosmic scale factor for distances. R is not the distance from any point in the universe to any other point but the number that determines the scale of actual distances in the universe. Just as all distances between points on the surface of the earth would double if the radius of the earth were doubled, so all distances between galaxies in space would double if R were doubled. The rate of expansion of the universe may therefore be expressed in terms of R, which, although not a directly observable and measurable quantity, can be expressed in terms of observables.

Picture now the three-dimensional space of the universe as a surface expanding into a fourth dimension and apply to it the concept of the speed of escape. The speed of escape of this expanding surface would then be expressed in terms of M and R, just as for the earth. If only Newtonian gravitational theory were involved, one could then write

$$V^2_{\text{escape}} = \frac{2M}{R}\,G.$$

Here the word *escape* attached to V as a subscript means one is not talking about the actual speed of expansion of the universe but about how large the speed of expansion would have to be for the universe just to keep on expanding. If the speed of expansion is just equal to $\sqrt{(2M/R)}\,G$, the universe is a flat, infinite, euclidean universe that will expand forever. If the speed of expansion is larger than this value, the universe is curved but curved like a vast open hyperbolic dish, so that it is infinite and the expansion will go on forever. If the speed of expansion is smaller than the value calculated with the above formula, the universe is a finite, closed universe—a spherical universe—whose noneuclidean three-dimensional space is curved inwardly.

To compare the present speed of expansion with the speed of escape calculated from the formula, a correct mathematical expression for the speed of expansion is needed. This is given by the general theory of relativity applied to the universe. In fact, this theory gives the following formula for the speed of expansion:

$$V^2_{\text{expansion}} = \frac{2M}{R} G - kc^2,$$

where c is the speed of light and k can have only one of the three values $+1$, 0, -1. Except for the term $-kc^2$ on the right-hand side, this formula for the square of the speed of expansion is just like the formula for the speed of escape. If $k = 0$, the speed of expansion equals the speed of escape and an infinite, euclidean universe exists. If $k = -1$, the speed of expansion exceeds the speed of escape and the universe is an infinite, noneuclidean, hyperbolic structure. If $k = +1$, the speed of expansion is less than the speed of escape and the universe is a finite, noneuclidean, spherical structure.

The theory also shows that k is related in a simple way to the deceleration parameter, so that if the value of the deceleration parameter can be deduced from the observational data, then k can be calculated. This has been done, as already noted, and k has been found to be $+1$, so that the speed of expansion is less than the speed of escape. This means that, placing $k = +1$ in the above formula, the quantity $(2M/R)G$ for the universe must equal the quantity $V^2_{\text{expansion}} + c^2$. Since the size of the universe, R, and its mass, M, are not known, it is better to express this relationship in terms of a quantity that can be measured—namely, the density of matter in the universe. This quantity is just the number of grams of matter, on the average, in each cubic centimeter of space in the universe.

To see how the matter density comes into this analysis, write the above relationship as an equation

$$\frac{2M}{R}G = V^2_{\text{expansion}} + c^2.$$

In other words, there must be enough mass in our universe to make the quantity $(2M/R)G$ larger than the square of the speed of expansion. If every term in this equation is divided by R^2, one obtains

$$\frac{2M}{R^3}G = \left(\frac{V_{\text{expansion}}}{R}\right)^2 + \frac{c^2}{R^2}.$$

But the speed of expansion divided by R is just Hubble's constant H, and $2M/R^3$ gives—except for a numerical factor that is very nearly 8—the average matter density, so that the equation says that in a closed, finite, expanding universe, a spherical universe whose rate of expansion is smaller than the speed of escape, the average density of matter multiplied by the gravitational constant G must be larger than one-eighth of the square of the Hubble constant. This, then, is a condition on the average density of matter in the universe, and hence on the total mass in the universe, if the universe is a closed spherical universe, as demanded by the observed data. Since Hubble's constant is known and the average density of matter can be ascertained from the knowledge of the number of stars in each galaxy and the number of galaxies in a given volume of space, this condition on the density can be checked. The observed value of Hubble's constant reveals that this condition demands that the average density of matter in the universe be larger than one gram of matter spread uniformly throughout a cube each of whose dimensions is about 35,000 kilometers. This is also the density one would get if one divided a hydrogen atom into 100,000 equal parts and put each part in a single cubic centimeter of space.

This is a clear-cut deduction from the theory that can be checked against observations. Since the observed numerical value of the deceleration parameter already shows that the speed of expansion of the universe is less than the speed of escape and, therefore, that the universe is finite and closed, the actual average density of matter in the universe should be found to be equal to, or larger than, the value stated above. The actual measurements of the density of mass in the universe, however, are disappointing; the observed density, based on the known number of stars and the estimated matter between the stars, is smaller by a factor of almost 100 than the density demanded by the known value of the Hubble constant. How is one to interpret this? If the observational data, the observed value of the deceleration, are accepted as showing that the expansion of the universe will not go on forever but will stop at some future time, the conclusion is that the density of matter in the universe is about 100 times greater than the density deduced from the presently known amount of matter in the observable universe. But where is this "missing matter"? Thus far, nobody knows; astronomers have looked for it everywhere for the last 20 years without success, although many proposals have been put forward.

Note that the amount of matter that is actually observed in such large aggregates of bodies as clusters of galaxies is too small by a factor of about 100 to account for the gravitational force that keeps these clusters together. Thus, the Virgo cluster of galaxies, which is about 40 million light-years away, contains some 2,500, which can be observed and counted on photographic plates. But these galaxies are moving about so rapidly within the cluster that about 100 times as much mass as is actually contained within these 2,500 galaxies is required to supply the gravitational pull needed to keep the cluster from dispersing. Indeed, it is difficult to see how such an aggregate of 2,500 galaxies moving about in the way they do could ever have collected into a cluster without the presence of a great deal more mass than that in the stars and the interstellar matter in these galaxies. In short, the gravitational glue that, in line with observations, is required to keep the universe from blowing up has yet to be found.

Evolving Universe Versus Steady-State Universe

Although the analysis of the red shifts strongly suggests a picture of a closed, finite, evolving universe, it has been questioned by various astronomers and cosmologists who have proposed what is called the steady-state, or continuous-creation, cosmology, which envisions an unchanging universe that had no beginning and will have no end.

The steady-state model of the universe was first proposed in 1948 by the British astronomer Herman Bondi and the American astronomer Thomas Gold on the basis of what they called the "perfect cosmological principle." Later, the British astronomer Fred Hoyle derived the basic features of the steady-state model by adding a term to Einstein's gravitational field equations. The "cosmological principle" of Einstein states that the universe must look the same to all observers no matter where they are and in what direction they are looking; Hoyle's "perfect cosmological principle" states that the universe looks the same at all times, past and future. The latter gives a steady-state model of the universe, since it says that things in the universe, from a large-scale point of view, must remain the same. Such a model precludes large-scale changes in the universe, although local changes, such as the evolution of life on the earth or the evolution of stars, are not forbidden.

If one accepts the perfect cosmological principle, and hence the steady-state cosmology, one must also accept the continuous creation of matter. The reason for this is that the recession of the galaxies means that the distances between the galaxies must be increasing, which cannot be allowed in a steady-state cosmology. To prevent this the steady-state cosmology requires the continuous creation of matter and the formation of new galaxies from this matter in order to keep the number of galaxies in a given volume of space the same at all times. In fact, the rate at which new galaxies must be formed in a given volume of space to keep everything looking the same at all times is three times the Hubble constant for each existing galaxy in the given volume. Put somewhat differently, the rate at which matter must be created on the average is one proton (or neutron) per year in each 10 cubic kilometers of empty space.

What kind of observations enable one to decide unambiguously between the steady-state cosmology and an expanding, evolving model of the universe? As noted, in deciding this question a great deal of weight must be given to the value of the deceleration parameter that is consistent with a closed, expanding universe, and hence an evolving one, but the missing mass presents a problem. The way to settle this matter definitely is to compare the concentration (the number per unit volume of space) of galaxies and other celestial objects a few billion years ago with the present concentration. If it were found that the galaxies in the universe were more closely crowded together some billions of years ago than they are now, it would have to be concluded that the radius of the universe—and hence the total available space—has increased steadily for billions of years while the total amount of matter has remained the same. This would prove that the universe has not remained the same but has been evolving, and the steady-state cosmology, with its continuous creation of matter, would have to be rejected. This is what the evidence shows.

To compare the concentration of galaxies in the universe a few billion years ago with the present concentration, one must count galaxies with large Doppler red shifts in a given volume of space and compare that count with the count of galaxies with small Doppler shifts in an equal volume of space. The galaxies having large Doppler red shifts are receding very rapidly and are at great distances from earth; hence, they show the universe as it was a long time ago.

Counts of these galaxies therefore reveal the concentration of galaxies when the universe was young. Counts of galaxies with small red shifts, on the other hand, give us the concentration of galaxies in the present state of the universe. Although it is difficult to obtain accurate galaxy counts at great distances because the smaller galaxies are too faint to detect with optical telescopes, enough evidence has been obtained during the last half century to show that the concentration of galaxies was greater a few billion years ago than it is now. From this it is deduced that the universe is not in a steady state but is evolving. As the universe expands and grows older the matter in it is thinning out; no new matter is being created spontaneously to keep the density of matter in space constant as the volume of space increases.

Evidence from Radio Galaxies

Although the counts of distant galaxies support the evolutionary model of the universe, the evidence from these counts is not conclusive enough to permit one to say definitely that the universe is not in a steady state but is evolving. But additional evidence has been obtained from radio signals from distant radio sources, and from the distribution of the amazing quasars, which indicates quite conclusively that the evolving model of the universe is correct.

Radio astronomy is a relatively young branch of astronomy; it was born quite accidentally in 1931 when the American scientist Karl G. Jansky, a radio engineer for Bell Telephone Laboratories, was investigating the unwanted and disturbing static that interfered with radio communication between America and Europe. Although most of this static arose from various kinds of electromagnetic phenomena on the earth, such as thunderstorms, he found that one component of the static, a steady hissing noise in his receivers, came from the fixed direction of the Milky Way. Indeed, Jansky demonstrated that the radio waves that caused the steady hissing in his receivers were coming from the direction of the center of the Galaxy. Thus was opened a new electromagnetic channel to outer space. In 1936 Grote Reber, an American radio engineer, built the first astronomical radio antenna designed specifically to receive cosmic radio signals. Since then, radio astronomical telescopes of various designs have been built all over the world, and radio signals have been received from the sun, the planets, stars of various kinds, galaxies, and the medium between. Of special interest here are those from certain galaxies that emit unusually large amounts of radio waves. Thousands of discrete radio sources that are too faint to be seen optically have been detected at great distances from the Milky Way, and the data obtained from these sources indicate that they are very distant galaxies with certain peculiar features. That they are at great distances is indicated by our inability to see them even with powerful optical instruments like the 200-inch telescope at Mt. Palomar; that they are peculiar is indicated by the very intense radio waves they are emitting, as compared to the relatively weak radio waves emitted by ordinary galaxies like the Milky Way and the spiral in Andromeda. Since these "radio galaxies" are at great distances from earth, their concentrations in space, as deduced from their numbers in various regions of the sky, reveal how compact the universe was billions of years ago, as compared to the compactness of the universe now. Since radio telescopes are far more sensitive than optical telescopes, the evidence from counts of radio galaxies is more reliable than the optical counts of distant galaxies. This strongly supports a closed—that is, finite—universe that is evolving as it expands; the number of radio galaxies per given volume of space at great distances (the compactness of matter in the universe billions of years ago) is larger than the number of galaxies per given volume of space at relatively small distances (the compactness of matter in the universe now).

The Amazing Quasars

Consider now the evidence presented by a group of very remarkable objects, the quasi-stellar objects or quasars, which look just like ordinary stars on photographic plates but which differ from stars in the following ways: (1) they are powerful sources of radio waves; (2) they emit great quantities of ultraviolet and infrared radiation; (3) they have exceedingly large red shifts—much larger, in fact, than the red shifts of any known galaxies or cluster of galaxies.

Quasars, or QSO's, for "quasi-stellar objects," were discovered quite accidentally in the early 1960s when astronomers noted that an object that appeared to be a typical star on a photographic plate was emitting radio waves and ultraviolet radiation with far greater intensity than a typical star. This immediately aroused the curiosity of astronomers in general, who began a thorough investigation of this strange "star," designated as 3C48 in the Cambridge Catalogue of celestial radio objects, with such unstarlike behavior. They immediately found another curious property that distinguishes this object from stars: its spectrum contains only a few strange broad emission lines instead of an abundance of familiar sharp, dark absorption lines. Emission lines are the bright lines emitted by the atoms in a hot gas; absorption lines are the dark lines formed in a continuous spectrum when atoms in a cool gas absorb certain colors from the light passing through the gas.

These emission lines stimulated a great deal of excitement because they seemed to have no relationship to the spectral lines of any known atoms and could not be identified at first. But the bold surmise by the American astronomer Maarten Schmidt at the Mt. Palomar Observatory that the lines belong to the hydrogen atom led to their identification and to the most surprising result of all. Schmidt, working with the quasar 3C273, showed conclusively that the observed emission lines in this quasar's spectrum are the usual visible spectral lines of hydrogen (the Balmer lines) but greatly red-shifted into a part of the hydrogen spectrum that made them unfamiliar. In other quasars it was found that the ultraviolet lines of hydrogen are shifted into the visible part of the spectrum. Atomic theory permits us to calculate the lines in the spectrum of hydrogen, and such calculations show, in agreement with observations, that hydrogen atoms emit various series of spectral lines, among which is a series in the visible part of the spectrum (the Balmer lines) and a series (the Lyman series) in the ultraviolet part of the spectrum. These ultraviolet lines are invisible to the naked eye but can be easily detected by ultraviolet detectors and arranged in a numerical sequence in order of wavelength. But if these lines are invisible to the naked eye when emitted by hydrogen atoms in earthly laboratories, how does it happen that they can be seen when the light coming to earth from quasars is analyzed? The answer is given by the Doppler effect, and this answer shows the quasars to be extremely unusual objects. The quasars are receding from earth at such high speeds that not only are the visible Balmer lines shifted far toward the red, but the ultraviolet lines of their hydrogen atoms are red-shifted into the visible part of the spectrum. This is such an enormous Doppler shift that the speeds of recession of quasars exceed by factors of 2 or 3 those of the most distant galaxies that were then known. This led many astronomers to conclude that the quasars are at far greater distances than the most distant galaxies known, a conclusion that brought with it some amazing consequences.

If observed quasars are at such vast distances (up to 10 billion light-years in some cases) as deduced from Hubble's law (the greater the red shift, the greater the distance), then they must be quite extraordinary objects and incredibly luminous to show up among stars on ordinary photographic plates; no ordinary galaxies at such vast distances have ever been photographed, because they are not luminous enough. If these cosmological distances for the quasars are accepted, they are clearly the most luminous objects known. Simple calculations show that to appear like a typical star on a photographic plate that has not been exposed very long, quasars such as 3C147 must emit about 100 times as much visible radiation every second as a typical galaxy like the Milky Way, with its 100 billion stars. From what kind of source do such vast quantities of energy come? This question has stumped astronomers since the discovery of the quasars and has led some physicists and astronomers to propose the hypothesis that quasars are not at cosmological distances from earth but, at most, a few hundred million light-years away. This, if correct, immediately leads to much smaller and manageable luminosities for quasars, but it presents difficulties and problems that are even more severe than that of the luminosities. To account for the very high observed speeds of recession of quasars, the proponents of this hypothesis suggest that quasars were hurled out bodily from the exploding centers of galaxies near them. But this raises some very serious difficulties, one of them being that some of the quasars ought then to be coming toward earth, since explosions throw material out in all directions, but not a single one is (there are no blue shifts). One is faced with the unanswerable question of how it is possible for an explosion in the center of the galaxy to propel a single massive object like a quasar in a single direction at a speed close to that of light.

One might try to solve the problem associated with the superluminosity of quasars by proposing that quasars are supergalaxies, many times larger than ordinary galaxies, but this will not work, because quasars are variable objects. Their luminosities do not remain constant but change periodically as though they were pulsating. This places an upper limit on their sizes. Some quasars are known to go through a complete cycle of variations once a week, which means that the diameters of such quasars cannot be larger than the distance light travels in a week (one light-week). This is extremely small compared to the diameters of ordinary galaxies, which equal tens of thousands of light-years. Thus, quasars are considerably larger than stars but tiny compared to galaxies. But one is left with the unsolved problem of their incredible luminosities.

What do quasars reveal about the evolution of the universe? Accepting the interpretation that their very large Doppler red shifts mean that they are at very great distances from earth, one must conclude that these objects present a very early stage in the evolution of the universe, anywhere from 4 to 7 billion years ago. Quasar counts should therefore enable one to compare the compactness of matter in the universe a long time ago with the present compactness. These counts show quite conclusively that the universe was much more compact billions of years ago than it is now and that it has, indeed, been evolving as the result of its expansion during the last few billions of years. This quasar-count evidence strongly supports the evolving models of the universe and practically eliminates the steady-state models; but there is still stronger evidence than this in support of the evolving models, in the background cosmic radio waves that permeate all of space and strike the earth from all directions.

Cosmic Background Radiation

In 1965 Arno A. Penzias and Robert W. Wilson, two American physicists at Bell Telephone Laboratories, published a paper in the *Astrophysical Journal* with the rather unappealing and unrevealing title "A Measurement of Excess Antennae Temperature at 4080MHz"; it would have attracted very little attention had it not been for a companion paper that appeared with it. In this companion piece, the American astrophysicists Robert H. Dicke, P. J. E. Peebles,

Peter G. Roll, and David T. Wilkinson explained the fundamental importance of these measurements for cosmology. Using the large radio horn antenna at Holmdel, New Jersey, that had been built to track the Echo satellites, Penzias and Wilson had detected a weak background radio signal that came equally from all directions in space at a wavelength of 7.35 centimeters, a phenomenon that they could not then explain. They called Dicke and his colleagues at Princeton about their unexplained receiver noise because the Princeton group was building a special radio telescope to look for the residual background thermal radiation that would still be around—in a greatly altered form—if the present universe had evolved from a very hot, compressed initial state (the initial "big bang" or "cosmic fireball") some 13 billion years ago, when the temperature was 100 billion degrees or more. Various theoretical physicists—the Russian George Gamow (1904–1970) in the late 1940s and the Princeton group, led by Dicke, in 1965—had predicted that if the universe had a very hot beginning, the very intense, high-frequency (short wavelength) gamma radiation that was then present and that dominated the initial state of the universe would still be present in an enormously red-shifted form, owing to the continual expansion of the universe that followed the big bang. Calculations show that the expansion of the universe must have increased the wavelength of the initial hot radiation by about a factor of 1,500, giving it just exactly the characteristics that the background radiation discovered by Penzias and Wilson has. Dicke and his colleagues explained all this in their paper in the *Astrophysical Journal*, and today most, although not all, cosmologists and astronomers believe that the homogeneous, isotropic background microwave radiation discovered by Penzias and Wilson is the greatly altered thermal radiation that filled the universe initially.

The importance of this discovery cannot be overemphasized, for it gives a clear and definite picture of how the universe began and how it evolved from its very hot birth to its present cool state. In its contribution to cosmology the discovery of the background microwave radiation ranks with Hubble's discovery of the relationship between the red shifts of the galaxies and their distances.

4 The Fireball Stage and the Beginning of Time

When radiation and matter are mixed together in an enclosed region, as in a hot furnace or in the deep interior of a star like the sun, the matter and radiation can be in equilibrium with each other in the sense that as the radiation flows back and forth in the confined space, it is constantly being absorbed and reemitted by the enclosed matter, so that the temperature in the enclosed region remains the same and the quality of the radiation—the mixture of colors or wavelengths in it—remains the same. Such radiation is referred to as thermal radiation; its temperature is that of the enclosed region. Strictly speaking, true thermal equilibrium between radiation and matter does not exist inside a star or in a furnace, because radiation is constantly seeping out of the star or furnace, so that the temperature within it tends to drop and would drop if the supply of radiation were not constantly replenished. Inside the furnace there is a continuous supply of radiation because fuel is burning, and inside the star nuclear fusion supplies the energy that leaves the star at its surface. If an impenetrable wall—impenetrable to the most penetrating radiation known—could be built around the star or furnace so that no radiation could escape, a state of true thermal equilibrium between radiation and matter at a definite temperature would be reached and the temperature of the star or furnace would remain constant. We may picture the initial compact state of the universe as being in precisely such a state of equilibrium because not only was all matter and

73

radiation concentrated in the small primordial sphere described above but so also was all of space, so that radiation could not escape, because there was nowhere for it to go. In a sense this initial state was a very hot black hole from which the present universe emerged.

Reconstructing the Past from the Present

Since there exists no direct observational evidence about the physical properties of the initial state of the universe, its properties must be deduced from the present state of the universe. This can be done by applying the laws of physics to the matter now moving about in the universe and by tracing things back to their origin in the past. Different initial conditions can be reached in this way, depending upon how the laws are applied, but the differences will be differences in certain details rather than in the overall initial picture. The laws of nature, as understood now, force one along a fairly well-defined path and therefore allow little leeway in this backward trip in time. This sort of reconstruction of the past is not unusual in science, for the same laws that permit prediction of the future also permit deduction of the past up to a certain point. If it is known that event A causes event B, it is safe to assume that the occurrence of B now means that A occurred in the past. Here caution in analysis is necessary because event B might stem from a number of different causes; but generally such multiple causes can be taken into account by working statistically with collections of different causes and effects. When this is done, slight variations that could be important if one wanted to establish a precise and detailed correlation between every event in the universe now and every event in the past are ironed out, so that what is left are broad statistical correlations, which are all that is of interest here.

An example of how this works will illustrate what is meant by statistical, rather than detailed, deductions of past events from present events: If one enters a warm room from the cold and observes some glowing embers in a fireplace at one end of the room, which has no other source of heat, one deduces from the warmth in the room and the few dying embers that at some moment in the immediate past the temperature in the fireplace was very high and the temperature in the rest of the room was very much lower. One further concludes, correctly, that many of the molecules moving about rapidly in the air were not free particles but parts of dense bits of matter (wood), as were the glowing embers and the warm ashes in the fireplace. These deductions give an accurate overall statistical picture of past events in the room that led to the present state of the room but not of every single detail of the past. Thus, it is impossible ever to discover where every single molecule now in the room was when the wood in the fireplace burst into flame or how each of these molecules was moving at that moment. But even if one could deduce the position and motion of each molecule in the past, one would still not have a completely detailed picture of the past, because one still would not know where each atom was in every molecule or where the single electrons were in the individual atoms. Fortunately, such detailed information is not only not necessary for reconstructing a picture of the past but is actually confusing. The detailed motions and positions of individual molecules are not important in the reconstruction of the past from the present; it is enough to know the earlier motions and positions of large numbers of molecules, which can be deduced unambiguously from present events. The exact history of any particular molecule is not important.

It is clear that if one has a good deal of unambiguous data about the present state of the universe, one can construct a reasonable picture of the initial state of the universe just by using the known laws of nature. Some may object to this, arguing that the laws of nature themselves may have been different billions of years ago from what they are now, so that it is incorrect to use today's laws to reconstruct the past. They may also argue that even if the laws were the same in the past as they are now, the constants of nature—the speed of light, the universal gravitational constant, the Planck constant of action—may have been different in the distant past. Such objections are really without merit because one has no choice in the matter; one must work with the laws and constants as one knows them. If it turns out that the laws and constants were different when the universe was much younger, this will show up in the attempt to reconstruct the past. Indeed, if the laws and constants of nature are, and have been, changing, the process of change is itself governed by a law that can be discovered by using presently known theories to reveal the initial state of the universe.

Unlike the very hot, homogeneous initial state of the universe of some 13 billion years ago, which may be described as a state of thermal equilibrium, the present state of the universe is not in thermal equilibrium. It is not at the same temperature everywhere throughout but consists of a mixture of very hot bodies, the stars, spaced at great distances and cool radiation and matter in the vast spaces between the stars. Wherever one looks, the universe appears to be running down instead of in a state of equilibrium. If one accepts the evidence presented by the continuous flow of radiation from the individual stars themselves—disregarding the expansion of the universe for the moment—it appears that a steady state of thermal equilibrium at a very low temperature will be reached at some distant time in the future when all the nuclear fuel in the universe has been exhausted and all the stars have become cold, inert objects. This could not have been the state at any time in the past, since the very existence of hot objects, such as stars, inside of which thermonuclear reactions are still occurring means that these objects must have been formed from a medium in which nuclear fuel was very plentiful. The existence of hot stars generating energy by means of thermonuclear transformations indicates, then, a previous state of the universe in which no stars were present but in which the physical conditions were proper for the formation of stars and in which hardly any thermonuclear transformations had occurred, so that the matter then was almost entirely in the form of hydrogen and helium.

To deduce the state of the matter in the universe before stars were formed one must now take into account the expansion of the universe. The expansion of the universe results in a continual cooling off of the entire universe just as a compressed gas in a container cools off when the gas pushes against a piston against which there is an external pressure. The gas cools off because it uses its own heat energy to do the work that is required to push out the piston. In the case of the universe, the expanding matter does work because it pushes outward against the total inward pull of gravity, which resists the expansion. Because the heat energy of the universe is thus used to keep the expansion going, the temperature of the universe drops.

If events are traced backward in time, it becomes apparent that the entire universe must have been much hotter in the distant past than it is now, because the universe had to use its own heat energy to bring it from a hot, contracted state to its present cold, expanded state. As the heat energy of the universe was expended to push the galaxies further away from each other against the force of gravity, which resisted this dispersion, the temperature of the universe fell. If one goes back just 1 or 2 billion years, one finds a somewhat more compact, warmer universe but one that does not look much different from the present universe. The galaxies are closer to each other, there are more quasars around, and there are fewer hot blue and blue white stars in the galaxies, but on the whole, things look familiar. The background cosmic radiation is somewhat bluer, and the overall intergalactic temperature is somewhat higher but not enough to make much difference in the relationship between matter and radiation. The dynamical properties of the universe are still dominated by the matter in the universe, with radiation playing a negligible role.

But the situation was vastly different some 10 or 11 billion years ago when the overall temperature of the universe was about 5,000°K (the absolute, or Kelvin, degree is the same as the centigrade degree, but the zero of the absolute temperature scale is 273° below 0°C). The radius of the universe was then less than a thousandth of its present radius, and this marked a crucial point in the history of the universe insofar as the formation of galaxies and stars is concerned. The attractive gravitational property of matter at that point began to dominate the dispersive property of the hot radiation, and galaxies and stars, as known now, began to form. In a sense, radiation and matter were no longer bound to each other in an influential way; radiation and matter went their own ways from that moment on.

The Initial Radius and Temperature

Before the universe reached this important point in its history, it had already passed through a series of states of extremely high temperature, with the temperature increasing very rapidly as one goes backward in time. It can be shown from very general thermodynamical principles that the absolute temperature of an expanding or contracting universe, consisting of a mixture of radiation and matter, varies inversely as the radius of the universe. This means that the absolute temperature of the universe decreased by a factor of 2 every time the radius of the universe doubled as it expanded from its early, very compact stage to the size it attained some 10 billion years ago, when its overall temperature was about 5,000°. When the radius of the universe was about one-billionth of its present radius, the overall temperature of the universe was about 5 billion degrees, and so on back to the beginning. The question that naturally arises here is, How far back does one have to go to reach the beginning and what were the conditions in the universe then?

Although no precise answer to this question can be given at the present time, there are various clues that indicate how long ago the zero moment of time occurred. To begin with, there are the data given by the receding galaxies, in the form of the Hubble constant and the deceleration parameter. Although there is still some uncertainty in the knowledge of these constants, they indicate an age of about 12 or 13 billion years. This means that the zero moment occurred some 12 or 13 billion years ago, when the radius of the universe was about 1,000 times the sun's present radius. In other words, all the mass and space of the universe was packed into a three-dimensional spherical shell whose four-dimensional radius was about 1 billion kilometers.

Another way of deducing this—or at least of assuring that these numbers are reasonable and, hence, not far from the correct values—is to consider the state of the radiation and the matter in the universe some 12 or 13 billion years ago. That the overall temperature in the universe depends inversely on the radius of the universe is written as follows: $T \propto 1/R$, where T is the absolute temperature, the symbol $\propto$ means *varies as,* and R is the radius. From this is deduced that the absolute temperature of the universe at this initial moment was about a trillion degrees. To see what conclusions can be drawn from this about the state of the radiation and the matter in the universe, another thermodynamic principle is introduced which tells us that the density of radiation in an enclosed region (the amount of radiation energy per cubic centimeter) varies as the fourth power of the absolute temperature. This says that if there are two enclosures—for example, two furnaces—containing radiation and the temperature of one furnace is twice that of the other, the amount of radiant energy in each cubic centimeter of the hotter furnace is 16 times as great as that in the cooler furnace. If the temperature of the hotter furnace were 3 times that of the cooler furnace, it would contain 81 times as much radiation in each of its cubic centimeters as the cooler furnace, and so on. This is also true for the whole universe; as its temperature rises (or drops), the density of its radiant energy rises or drops by the fourth power of its temperature change. But, as already noted, the temperature of the universe varies inversely as the radius of the universe. Hence, the density of radiant energy in the universe varies inversely as the fourth power of the radius of the universe. On the other hand, the density of matter, which is defined as mass divided by volume, in the universe varies inversely as the volume of the universe, which itself varies as the cube of the radius. Thus, the density of matter in the universe varies inversely as the cube, not as the fourth power, of the radius and hence varies more slowly with the radius than the density of the radiant energy. As the radius of the universe changes, the effect of this change on the density of radiation is more drastic than on the density of matter.

At the present time the density of the radiant energy (the microwave background) in the universe is about 100,000 times smaller than the density of matter, so that man is in a matter-dominated era; but things were quite different in the distant past. Some 10 billion years ago, when the universe was about 2 billion years old, more or less, and the temperature was about 5,000°, the density of radiant energy was about equal to the matter density, but before that, radiant energy dominated over matter. At a temperature of 1 trillion degrees or more the density of radiation in the universe was about 100 trillion grams per cubic centimeter, whereas the density of matter, in the form of particles such as protons and neutrons, was about 10 million grams per cubic centimeter, so that the density of the radiant energy at the initial or zero moment of the universe was some 10 million times as great as that of matter. Under these conditions, matter as it is now known could not exist; not only would this intense radiation rip all atoms apart, but it would also tear the nuclei of all the atoms into shreds, so that only basic particles like protons, neutrons, electrons, neutrinos, and the corresponding antiparticles could exist in equilibrium with this radiation. Antimatter would exist at a temperature of a trillion degrees because a good deal of the radiation would change itself into many pairs of particles and antiparticles in accordance with Einstein's famous relationship between energy and mass, $E = mc^2$. This equation tells us that the amount of energy E that is equivalent to a given amount of mass m, and hence the amount of energy released when mass is annihilated, equals the mass m multiplied by the square of the speed of light. This is a two-way relationship because it holds not only for the transformation of mass into energy (the annihilation of a particle by an antiparticle) but also for energy into mass, and this transformation of energy (radiation) into mass (particles and antiparticles) occurs when the temperature of the radiation is so high that the average energy of individual photons of the radiation (a photon is a quantum, or particle, of radiation) is equivalent to twice the mass of the particle being created in accordance with Einstein's energy-mass equation.

The Initial Radiation and Matter

Now consider how Einstein's mass-energy relationship leads us to a picture of the conditions in the initial, compact state of the universe when its temperature was about a trillion degrees. A simple thermodynamic formula permits the calculation of the average energy of the individual photons in the very hot, energetic radiation that dominated the universe at that time; it is found that, on the average, the energy in a single photon was equivalent to the mass of more than 1,000 electrons. This means that copious quantities of electrons and positrons (positrons are antielectrons) were produced from the radiation, and these then existed in equilibrium with the remaining radiation. Not all the hot radiation was transformed into electron-positron pairs but only as much as was necessary to produce a state of equilibrium between the number of electron-positron pairs produced and the number of photons left over. Since each photon had enough energy to produce about 500 or 600 electron-positron pairs, the number of electrons and positrons that were present in the initial state of the universe greatly exceeded the number of neutrons and protons. No neutron-antineutron pairs or proton-antiproton pairs were produced from the radiation in the initial state, because the mass of a proton is about 2,000 times that of an electron, and this exceeds the total mass that can be produced on the average from the energy of an individual photon in trillion-degree radiation. The number of nucleons was thus just about the same then as it is now, so that neutrons and protons, which were present in equal numbers, constituted less than one percent of the total number of particles which were present in this initial state of the universe. In addition to vast numbers of electron-positron pairs, there were also numerous neutrino-antineutrino pairs and muon-antimuon pairs in equilibrium with the radiation. The muon is a very short-lived particle that is about 208 times as massive as the electron but is like the electron in all other respects.

One may wonder at this point how it is that in a very compact universe there is so much more energy packed into the radiation than is found in the weak, dilute radiation that now permeates all of space. The additional energy comes from the work done by the gravitational force when it compresses all the radiation and matter into a small sphere; it is essentially the potential energy of the gravitational field that is present when the universe is greatly expanded, as it is now. If the universe contracted down to a sphere one-trillionth of its present size, the gravitational potential energy of the universe would be transformed into increased kinetic energy of the nucleons and increased radiation energy. At first the kinetic energy of the nucleons increases more rapidly than the energy of the radiation, but after a certain point, the radiation gets most of the potential energy released by the universe as it contracts. The radiation thus acquires enough energy per photon to create electron-positron pairs.

Before the Big Bang

There is nothing in the laws of nature, as far as is known, that says that the universe could not have been still smaller and hotter than it was 13 billion years ago. If one tries to construct a picture of a still earlier state of the universe, one must go backward in time very slowly because a fraction of a second can make a big difference in conditions. If one could look at the universe as it might have been a ten-thousandth of a second before the zero moment, one would find it considerably hotter than 1 trillion degrees. In fact, if an earlier state of the universe did exist, it must have been so hot that protons and neutrons were themselves torn apart into their constituent particles whose existence various physicists have postulated. These basic particles, which physicists call quarks, partons, or unitons, and their antiparticles would then have been in equilibrium with the very hot photons, electrons, and other particles and their antiparticles. Since no one has yet succeeded in tearing a proton apart, even with the very large energies available in the largest accelerators, quarks or unitons, if they exist, must be very massive. But if the very hot radiation that existed in a universe whose temperature was well above a trillion degrees could have dissociated protons into very massive quarks or unitons, some of the quarks or unitons that were present initially could have failed to coalesce into protons and may still be around. If such particles are around and if they are massive enough—that is, of the order of one one-hundred-thousandth of a gram—they could supply the missing mass that is needed to keep the universe from expanding forever. Only one such quark or uniton for every 100,000 trillion ordinary protons is all that is required to account for the observed deceleration parameter—that is, for the missing mass. It is obvious that detecting such rare particles, even though they supply more than 99 percent of the total mass in the universe, is extremely difficult, but if such particles were discovered, their existence would clear up many obscure points about the initial, compact state of the universe. Until that time, uncertainty about the conditions in the initial state of the universe will persist.

Although no one is sure just how high the temperature of the universe was at the moment of the big bang, a reasonably complete picture of the evolution of the universe since then can be drawn. The very early stages after the explosion—say, during the first three seconds—were marked by a very rapid drop in temperature from 1 trillion degrees to about 5 billion degrees, a factor of 200. During these three seconds of rapid cooling, the muons and antimuons annihilated themselves and the neutron-proton balance began to shift in favor of protons. At the same time, the electron-positron pairs began to annihilate, and this went on until the universe consisted almost entirely of photons (gamma rays), neutrinos, and antineutrinos, with a smattering of electrons, neutrons, and protons, the number of electrons equaling the number of protons. This phase of the history of the universe was still dominated by intense radiation. Thus, at the end of the first three seconds of its life the universe was still very hot, but it had already cooled off sufficiently for most of the neutrons to have been changed to protons. In fact, at this point the neutron-proton ratio was 1 : 5. Photons, neutrinos, and antineutrinos dominated; the only electrons that were present were the relatively few that were released when the neutrons changed into protons.

During the next three minutes, the universe cooled off to a temperature of about a billion degrees and the neutrons then began to fuse with protons to form helium and possibly elements like lithium, beryllium, and boron. At the end of this stage, which may have lasted for a few more minutes, the universe consisted of 25 percent ionized helium (alpha particles) by weight, and 75 percent ionized hydrogen (free protons and electrons) by weight, with some traces of deuterium, lithium, beryllium, boron, and, possibly, carbon present. The continued expansion of the universe had by that time reduced its temperature well below a billion degrees, and the dominance of radiation began to give way to the dominance of matter. But matter did not really begin to dominate the universe until the temperature had dropped to a few thousand degrees, which occurred after about a billion years. By that time the free protons and alpha particles (helium nuclei) had all combined with the free electrons that were around to form the primordial hydrogen and helium from which the first stars (called population-II stars) were to be formed.

At this point it must be emphasized that no appreciable quantities of heavy elements, such as carbon, oxygen, neon, and sodium, could have been formed during this early history of the universe when it was very hot and dominated by radiation. High temperatures are required to build up the heavy elements from hydrogen and the early universe was certainly hot enough; but this buildup takes a very long time, and since the high-temperature phase of the universe was over in a matter of hours, there was not enough time for the thermonuclear fusion of hydrogen into heavy elements. Since there are no stable atomic nuclei with atomic weights 5 or 8, the buildup of elements heavier than helium cannot proceed directly by the thermonuclear fusion of a proton with an alpha particle (the helium nucleus) or by the capture of a neutron by the helium nucleus. (The atomic weight of an element is its weight on a scale on which the oxygen atom has the weight 16; thus, the atomic weight of helium is 4; of hydrogen, 1; and of carbon, 12.) Such processes do not occur because they would lead to a nucleus of atomic weight 5, which does not exist. For the same reason, two alpha particles (two helium nuclei each of atomic weight 4) cannot combine to form a heavier nucleus of atomic weight 8 because no such

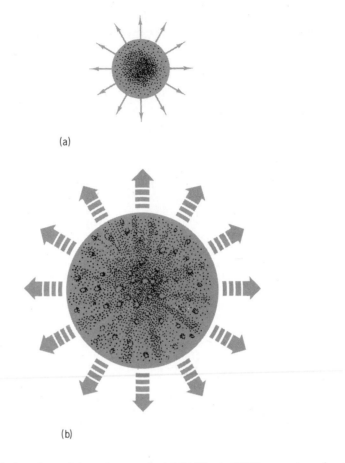

(a)

(b)

(a) The radiation phase of the universe—the initial "fire ball." Temperature about 10^{11} °K completely dominated by intense high-energy gamma rays.

(b) The matter phase of the universe. Galaxies being formed; the gravitational force begins to take control and slow down the expansion considerably.

nucleus exists. In other words, the atomic weight gaps at 5 and 8 cannot be bridged by the direct capture of neutrons and protons or by the thermonuclear fusion of helium nuclei.

There is a way this can be done, however, but it requires a high density of helium nuclei as well as a high temperature. If enough alpha particles are present in a unit volume at a temperature of a few hundred million degrees, enough pairs of helium nuclei will collide violently enough every second to form small amounts of the unstable nucleus of atomic weight 8, an isotope of beryllium. Ordinarily these unstable beryllium nuclei would immediately break down again into two alpha particles, but if the density of alpha particles is high enough, some of these unstable beryllium isotopes will capture a third helium nucleus to form the ordinary carbon nucleus. In this way the nuclear gaps at atomic weight 5 and 8 can be bridged. But for this to happen at a reasonable rate—which must have occurred somewhere under the appropriate conditions if the presently observed abundances of carbon and other heavy elements are to be accounted for—the concentration of helium nuclei must be very high. This was not the case in the early universe when the temperature was about one billion degrees and therefore high enough to produce this triple helium reaction. At that point in the life of the universe the density of helium was far too low for more than a relatively few helium nuclei to be fused into carbon by this triple alpha process.

If the temperature of the universe had remained at a few hundred million degrees for a billion years or more, an appreciable amount of carbon and even heavier elements might have been built up before stars were formed, but the universe remained at this temperature for no more than a few hours. One would then have required incredibly high helium concentrations for this carbon buildup to occur. Since the concentration of helium in the early universe was not high, one must look elsewhere for the buildup of the heavy elements. One does not have far to look, because the deep interiors of the earliest stars that were formed in the universe passed through a very high temperature and high density stage during which conditions were just right for the thermonuclear fusion of helium into carbon via the triple alpha process. Since this stage in the lives of these stars lasted for a billion years or more, there was ample time for the thermonuclear buildup of carbon.

The Formation of the Galaxies

As noted, after its initial outburst, the universe cooled off very rapidly as it expanded, and passed from radiation dominance to matter dominance after about a billion years, by which time the temperature had dropped to about 5,000°. At that point, the radius of the universe was less than one-thousandth of its present value, but conditions were already set for the formation of vast gaseous clouds of neutral hydrogen and helium. These clouds were the precursors of the present galaxies and globular clusters, but exactly how the galaxies of stars are formed from gas clouds is still very much of a mystery. It is necessary, then, to leave this question for the moment and concentrate on the problem of the formation of the gas clouds themselves.

In the very early and very hot stages of the universe, before the big bang occurred, the force of gravity, the weakest of the four known forces of nature, dominated and held the material of the compact universe together. The other forces had hardly any effect on matter at this stage because the temperature was then so high that the nuclear and electro-magnetic forces were ineffective in binding basic particles into known nuclear and atomic structures. How long the force of gravity held the primordial ball of matter together is not known, but once the big bang occurred, gravity could no longer keep things under control; however, it did begin to slow the expansion down immediately and finally cooled things off. The expansion initially was extremely violent, but the retarding effect of gravity, although hardly perceptible at first, in time reduced the temperature of the radiation and the speed of expansion to a point where the force of gravity could begin to organize matter into large structures. This happened after the expanding material had cooled down to a point where first the nuclear force and then the electromagnetic force did some preliminary work. The nuclear force transformed about 25 percent of the original neutrons and protons into helium when the temperature was still high and then, when the temperature had sunk to a few thousand degrees, the electromagnetic force organized the electrons, protons, and alpha particles into neutral hydrogen and helium atoms. From this point on, however, until the expanding material coalesced into individual stars, the nuclear force played no further role and the electromagnetic force played a minor role in organizing matter into structures.

The present state of affairs in the universe began with the force of gravity directing things after the universe had passed from the radiation-dominated stage to the matter-dominated stage. This is when and how processes that led to the present highly differentiated state of the material universe began. At that time, the universe was still quite homogeneous and undifferentiated, but the first step toward the present state was taken when the force of gravity initiated a hierarchy of condensations that began with the fragmentation of the initially uniform diffuse matter into huge clouds and ended with stars, planets, and complex organic molecules.

Although the force of gravity between any two particles, such as two atoms, is extremely weak, the total force between many such particles can become quite strong—in fact, strong enough to bind these particles into a single structure—if these particles get close enough together and are not moving at too great a speed. The general conditions that must be fulfilled in a diffuse gas for a collection of particles of the gas to stay bound together gravitationally can now be deduced. If such a gas were distributed uniformly throughout a vast region of space and the particles of the gas were moving about in a perfectly random way, the distribution would remain uniform and gravitational fragmentation would not occur. But complete and perfect randomness of motion does not occur; there is always a chance that more particles of gas will collect momentarily in some regions than in others. These local and random fluctuations in the density of the gas go on continuously, so that under the appropriate conditions the uniform diffuse distribution can—and, indeed, will—break up into large clumps of particles that stick together to form gravitationally bound structures. The appropriate conditions for a single random condensation to maintain itself as a gravitational structure within the diffuse distribution, instead of dispersing again, will occur if the total random condensation involves such a large number of particles and is distributed throughout such a small region of space that the force of gravity throughout the condensation is strong enough to prevent the individual particles from escaping.

Here is a problem similar to that of the speed of escape: if a body is thrown away from the surface of the earth fast enough, it will escape to infinity if its speed is larger than a certain value that can be derived from the mass and radius of the earth. The same thing applies to the particles within the condensation. Owing to the temperature of the gas in the condensation, the atoms—hydrogen and helium in this case—will be moving around with a certain average speed. In fact, the average speed of molecules or atoms in a gas is proportional to the square root of the absolute temperature. Hence, the higher the temperature, the greater the tendency of the condensation to disperse. On the other hand, the restraining force, the force of gravity, on a molecule or atom somewhere near the surface of the random condensation depends on the total mass of the condensation divided by its radius. From these two facts one can deduce a condition that must be fulfilled if a condensation is to achieve stability and maintain itself. This condition can be stated in a general way as follows: A condensation will survive as a gravitational structure if the total gravitational potential energy, the binding energy stemming from the gravitational attraction, exceeds, by a certain factor less than 2, the total energy of the thermal motion of the molecules or atoms in the condensation. From this condition, using a little algebra, one can obtain an inequality that tells how large a condensation must be before it will perpetuate itself as a gravitationally stable structure instead of breaking up and dispersing again. This inequality can be given either for the radius of the condensation or for its total mass and is expressed in terms of the temperature and the density of the condensation. The inequality may be expressed in terms of these two physical parameters because the energy of the thermal motion is given by the absolute temperature T and the gravitational potential energy can be expressed in terms of the density. The higher the temperature, the greater is the thermal kinetic energy and, hence, the greater is the tendency for the condensation to disperse. The greater the density, the greater is the gravitational binding, because more particles are squeezed together. When one takes these things into account, one finds that any given condensation will perpetuate itself gravitationally if the radius of the condensation, expressed in centimeters, is larger than 44 million multiplied by the square root of the temperature and divided by the square root of the density. This inequality is written in centimeters as follows:

$$R \geq 4.4 \times 10^7 \sqrt{\frac{T}{\rho}}$$

where R stands for radius, T for absolute temperature, and ρ for density.

If one considers the total mass M of the condensation instead of the radius, one finds that the mass, in grams, must exceed 40 billion trillion multiplied by the square root of the cube of the temperature and divided by the square root of the density if the condensation is to survive. This inequality is written in grams as follows:

$$M \geq 4 \times 10^{22} \sqrt{\frac{T^3}{\rho}}.$$

To see what these inequalities imply, one uses the first one to calculate how large the condensation in the earth's atmosphere would have to be for the condensation to exist as a compact ball of air, separate from the rest of the atmosphere. Since the mean temperature of the atmosphere is about 300°K and its mean density is about one-thousandth of a gram per cubic centimeter, one finds on substituting these numbers on the right-hand side of the inequality for R that the radius of such a condensation would have to exceed 20 billion centimeters, or 200,000 kilometers, which is much larger than the earth's diameter. This shows why the earth's atmosphere does not fragment gravitationally into separate spheres of gas; the atmosphere is too thin.

One now applies the second inequality, which is expressed for the mass of the condensation instead of for its radius, to the gravitational fragmentation of the primordial gas in the expanding universe shortly after the temperature had dropped to a point where matter began to dominate. Fortunately, one does not have to insert the values of the temperature and density of the gas at that particular time. The present values of T and ρ can be used for the calculation because the cube of the temperature and the density change in exactly the same way as the universe expands. This means that the mass of a stable cloud formed by gravitational fragmentation must be the same regardless of when, during the expansion of the universe, the fragmentation occurred. If one substitutes numbers for T and ρ in the inequality for the mass M, one finds that the mass of a fragmentation must be at least 5 million times greater than the mass of the sun if it is to survive.

This number is the same for fragmentations that occurred a few billion years ago and for those that are occurring right now, but the sizes are not the same. A few billion years ago the fragmentations were small and very dense, whereas today the fragmentations are extremely large and very tenuous.

A mass of 5 million suns is considerably smaller than the mass of an average galaxy, so that these clouds cannot be considered forerunners of galaxies. But they may very well have been the precursors of the remarkable globular clusters of stars that form halos around the cores of galaxies. The geometry and the general physical properties of these ensembles of stars (a globular cluster contains anywhere from a few hundred thousand to a few million stars) are quite similar to the computed properties of the stable condensations described above. More than 100 such clusters have been observed revolving around the nucleus of the Galaxy, and it is known that these objects are extremely old because they contain some of the very oldest stars. Their shape, which is remarkably spherical, indicates that they are hardly spinning at all. Moreover, the masses, radii, and luminosities of globular clusters are about the same wherever they are found. This indicates that they were all born in pretty much the same manner and in about the same epoch in the history of the expanding universe. Since the globular clusters are not spinning, they must have evolved from gas clouds that had very little or no spin at all; such gas clouds would have resulted from gravitational fragmentation that occurred while the expanding primordial gas was still hot enough to be ionized. The reason for this is that a hot, ionized gas is very viscous, so that rotations and turbulences in general are damped out very quickly. Hence, spherical clouds formed in such a hot gas have little or no angular momentum and thus retain their spherical shape instead of becoming disklike, as they would if they were spinning rapidly. Such clouds thus have the dynamical and geometrical properties that one would expect to find in the precursors of globular clusters.

How then were the galaxies formed if the fragmentation condition outlined above led to rather small gravitationally bound clouds? There are various possibilities here. To begin with, the fragmentation condition stated above gives the smallest mass that a cloud can have if it is to survive as gravitationally bound, but it does not place an upper limit on the size or mass of such a cloud. In other words, very massive clouds, massive enough to form galaxies, can be formed during the gravitational fragmentation of a streaming fluid like the primordial gas, even under the conditions outlined above; but whether enough such large clouds were, or could actually have been, formed during this early period to account for the total number of galaxies that now exist is questionable.

A point, left out of the discussion above, that favors the formation of much larger clouds than those deduced above must now be considered. This point is the expansion of the universe, which introduces additional kinetic energy and therefore makes it more difficult for clouds of small mass to form; the introduction of the expansion kinetic energy upsets the relationship between the gravitational potential energy and the total kinetic energy established previously. When the expansion of the universe is taken into account, the minimum mass of a cloud that can be formed and can survive during gravitational fragmentation is found to be considerably larger than the value of such a mass deduced above without taking the expansion into account.

There is, however, another series of events that could have led, and probably did lead, to the formation of galaxies. These events must be considered because, in spite of the expansion of the universe, the general tendency during gravitational fragmentation was for clouds of gas to form with masses about equal to those of globular clusters. Even with the expansion of the universe taken into account, it appears that clouds with masses equal to, or larger than, the mass of a typical galaxy could have been formed only in the relatively short interval of time between the epoch when the temperature of the universe was about 10 million degrees and the epoch when the temperature had dropped down to about 4,000°, when protons and electrons were able to recombine to form neutral hydrogen. But this period of time was probably much too short for many gas clouds as large as galaxies to form, although some probably did. When the temperature had dropped below 4,000°, the gas clouds that were formed were about as massive as globular clusters, so one must ask whether galaxies could have been formed gravitationally from such clouds. Theoretical work done by Dicke and Peebles indicates that such gravitational accumulations of gas clouds are possible and probably did occur some 200 million years after the initial, fireball phase.

The figure 200 million years is important here because the expansion of the universe was so rapid when the gas clouds were first formed, and for about 200 million years after that, that the clouds moved apart in response to the expansion instead of coming together gravitationally. But in time, as the expansion slowed down, groups of clouds did begin to drift together because of their mutual gravitational attraction. This tendency was greatly accelerated when the universe was about 200 million years old because by that time the expansion of the universe had slowed down sufficiently to permit groups of thousands to millions of clouds resembling globular clusters to collect into single gravitationally bound structures as massive as galaxies. When these large numbers of clouds rushed together, violent collisions between them, with the release of vast quantities of energy, were inevitable and stable galaxy-like structures were thus formed. Not all the clouds that came together collided in this way; some of them remained gravitationally bound to the others, but on the outskirts of the galaxylike structures, and converted their gas into stars, thus becoming what are now called globular clusters. The globular clusters that are now seen may be considered as, in Dicke's words, the fossil remains of the original gas clouds.

The Evolution of the Galaxies

Although the high temperature of the primordial gas, the expansion of the universe, and the gravitational attraction between the clouds were undoubtedly the most important factors in the transition from a hot, homogeneous, expanding gas to a distribution of widely separated, discrete structures consisting of stars, dust, and gas, other phenomena played an appreciable role also. If all galaxies were about the same size and shape, the problem of accounting for their structure would be greatly simplified, but this is not so. They come in a variety of shapes, sizes, and content, ranging from almost perfectly spherical structures with no trace of dust or spiral arms, through highly developed spirals, to dust-laden irregular structures. Following the pioneering classification work of Hubble, astronomers divided galaxies into four well-defined categories: ellipticals, normal spirals, barred spirals, and irregulars. The ellipticals are beautifully symmetrical structures that show no trace of dust nor spiral arms; they are smooth spherical or ellipsoidal aggregates of red and yellow stars distributed about dense bright cores. No blue or blue-white stars are found in ellipticals. A spiral is a flattened, disklike structure with a luminous, reddish bulging core from which, generally, two well-defined spiral arms emerge. The spiral arms, which are characterized by dust and very luminous blue-white stars generally extend once or twice completely around the core. A barred spiral has a long bar that cuts right across the small luminous core, and two spiral arms—one from each end of the bar—encircle the galaxy. In the case of some barred spirals, the core itself has a small spiral structure. The irregular galaxies show no definite spiral or ellipsoidal structure; they are aggregates of dust, gas, and stars distributed in irregular patterns.

If galaxies were, indeed, formed from aggregates of large gas clouds that fragmented from the primordial gas and were brought together by their mutual gravitational attractions, one must try to understand why such a variety of structures exists now. One possible explanation is that when galaxies were first formed, they all had the same general shape and then evolved, at varying rates, to the present variety of shapes and structures. This idea of the evolution of galaxies was very attractive in the early years of galaxy research, particularly in the light of Hubble's classification, in which the galaxies were arranged in a diagram according to shape. In this scheme, the spherical and elliptical galaxies are arranged along a line with the most nearly spherical galaxies on the extreme left and the more highly elliptical ones to the right. The spirals are then arranged along two diverging diagonal lines to the right, with the normal spirals along one of these lines and the barred spirals along the other. This scheme suggests that galaxies began their lives as spherical structures and then flattened into ellipticals as they contracted gravitationally and increased their rotational speed; the ellipticals then evolved further developing either as normal or as barred spirals. The irregular galaxies were pictured as the very youngest galaxies, which would ultimately contract into ellipticals and then into spirals.

This evolutionary theory was quite popular for a while, but as more and more data were collected, it ran into insurmountable difficulties. An analysis of the rotation of the spherical galaxies shows, for example, that these galaxies are rotating much too slowly ever to flatten out into pronounced ellipticals. Moreover, ellipticals cannot develop spiral arms, because their rotation pattern is not of the differential kind, which is required for the formation of spiral arms. One more argument against this evolutionary scheme is that the masses of the ellipticals are larger than those of the spirals. One would thus have to introduce some mechanism that could get rid of mass as the galaxies change from ellipticals to spirals.

Some astronomers have proposed an evolutionary scheme that is just the reverse of that just described. They suggest that galaxies begin as irregular structures consisting of gas and dust from which stars are constantly born. As more and more stars are created from the gas and dust, they fall in toward the center, thus forming a core of the very oldest stars, and leave dust and gas in the outer regions. As the newly created stars collect toward the center, the speed of rotation of the galaxy steadily increases causing the regions of dust and gas outside the core to flatten. The core will also flatten but not nearly as

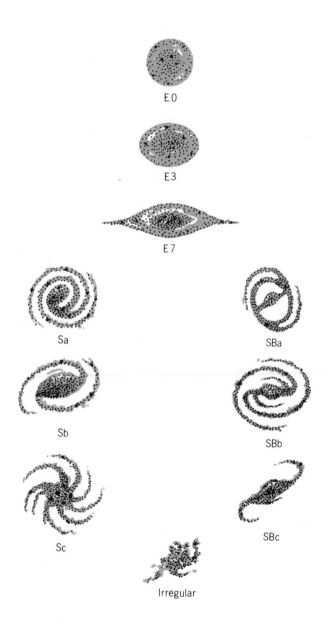

E 0

E 3

E 7

Sa

SBa

Sb

SBb

Sc

SBc

Irregular

The Hubble classification of galaxies. This scheme suggests that the galaxies evolve from non-spiral (elliptical) to spiral structures.

much as the outer regions. Since, because of friction, the dust and gas will not rotate as rapidly as the core, spiral arms will be formed by the gas and dust that lag behind the core. This is always the case when there is differential rotation in a region in which there are two different liquids. A good example of this is seen in the behavior of a bit of cream dropped on the surface of black coffee rotating in a cup. As the cream falls onto the surface of the spinning coffee, it lags behind the coffee and forms spiral arms. According to this theory, the spherical and elliptical galaxies are the end products of this process of transformation of dust and gas to stars; when all the gas and dust in the outer regions have been used up, the remaining structure is an elliptical galaxy.

The difficulty with this evolutionary theory is that there is no mechanism that can lead from a flattened structure, such as the core of a spiral galaxy, to a spherical or even an ellipsoidal structure. Moreover, as already noted, the masses of the ellipticals are, in general, larger than those of the spirals and there is no known way in which a spiral can become as massive as an elliptical.

Oort has proposed an evolutionary scheme in which the ellipticals form one evolutionary sequence and the spirals another without any crossover: ellipticals remain ellipticals and spirals remain spirals during their entire lives. According to this theory, which seems to have no objectionable features, the factors that determine whether clouds of gas and dust will ultimately begin their galactic lives as elliptical or spiral structures are the amount of dust and the net amount of rotational motion in the clouds. If an aggregate of gravitationally bound clouds is to become a spiral galaxy, it must start out with at least one percent of its material in the form of dust and with a fair amount of rotational motion. On the other hand, if a gravitational collection of clouds has no dust, or very little, and not very much rotational motion, it will ultimately become an elliptical galaxy.

This theory suggests that elliptical galaxies are older than the spirals because they must have been formed from clouds that consisted of the primordial hydrogen and helium only, with no dust present. Particles of dust in interstellar space cannot be formed from hydrogen molecules and helium atoms alone; heavier elements such as carbon, oxygen, and iron are required. But such heavy elements were not present when the primordial gas fragmented into clouds, so that these very earliest clouds were essentially hydrogen and helium gas clouds. Moreover, because these early clouds had very little rotational motion owing to the high viscosity of the primordial gas when they were formed, these ensembles of clouds contracted into smooth ellipsoidal structures, with no trace of dust or spiral arms, instead of into gas-laden, spiral structures. Owing to the high density of the gas in such a structure, when the constituent clouds had contracted down to an ellipsoid, most, if not all, of the gas condensed rather quickly into population-II stars, stars that consist mostly of hydrogen and helium with hardly any trace of metals in their atmospheres.

On the other hand, if the clouds that coalesced into galaxies contained an appreciable amount of dust and rotational motion, the rapid rotation resulted in a disklike structure with the dust in the outer regions rotating more slowly and trailing behind the rapidly rotating core, thus forming spiral arms. The dense gaseous material near, and in, the core condensed into stars quite rapidly (population-II stars). Thus, whereas elliptical galaxies consist of very old, reddish population-II stars throughout, spiral galaxies contain reddish and yellowish old population-II stars in their cores and young stars of all colors, particularly very luminous blue and blue-white stars, in their spiral arms. Irregular galaxies, which account for about 3 percent of all the brightest known galaxies in the sky, are probably among the youngest galaxies. Although they contain gas, dust, and stars of both populations, they show no rotational symmetry or any trace of spiral arms.

Certain curious features about spiral arms cannot be explained by appealing to rotation only. In the case of the coffee and the cream, one sees that the spiral arms formed by the cream wind around many times as the coffee rotates. In fact, every time the coffee spins completely around, the spiral arms are wound around again. There are thus numerous thin, closely spaced spiral arms. This is not so in spiral galaxies, which, in general, consist of two or three widely spaced spiral arms, in spite of the fact that these galaxies have made many complete rotations since their formation. Thus, the Milky Way, which rotates once every 250 million years, must have spun completely around at least 20 times since the solar system was formed, and yet it contains only three spiral arms instead of the 20 or more one would expect if the material in the spiral arms simply trailed the solidly rotating core. The solution to this problem is to be found in the coexistence of stars and dust in the spiral arms. As the stars revolve in fixed but different orbits around the core of the galaxy, the gas and dust fall into spiral patterns that are formed by density waves generated in the galaxy. One might compare this to the wave pattern formed by dust on a vibrating metal plate. When the plate is set vibrating, the dust arranges itself along the antinodes of the standing waves.

Another feature that is difficult to understand in terms of the rotation alone is the continued existence, in spite of the rotation, of certain elongated structures in galaxies. It is, for example, very surprising to find the long bars in the barred spirals. That such structures have existed for billions of years, instead of being wound up into spiral arms, means that these bars are spinning around like rigid bodies. The gas and stars at the ends of the bars must be revolving around the core much more rapidly than the material near the core. From this it follows that some force—probably magnetic fields—in addition to gravity must be operative. If this were not so, the stars in the outer regions of the bars would revolve around the core more slowly than the inner stars, just as the outer planets in the solar system revolve around the sun more slowly than do the inner planets, and bars could not exist.

In addition to the barred spirals, there are certain peculiar galaxies with what appear to be single rigid arms extending from the core. Here again such structures cannot be accounted for without the introduction of magnetic fields. That magnetic fields play an important role in the stability of the spiral arms has been shown by two physicists, the Italian-American Enrico Fermi (1901–1954) and the Indian Subrahmanyan Chandrasekhar, whose theoretical work in this area demonstrated that the magnetic fields tend to make spiral arms behave as though they were rigid. The magnetic fields in galaxies need not be strong for this purpose; in fact, Fermi and Chandrasekhar showed that the stability of spiral arms against lateral disruption can be maintained by magnetic fields that are weaker than the earth's magnetic field by a factor of about a million.

The Initiation of Star Formation in Galaxies

Without stars there could be no planets like the earth and, hence, no stable intelligent life. To understand the origin of the world, one must therefore begin by considering the origin of stars; indeed, the formation of planets like those in the solar system is an inevitable consequence of the formation of stars like the sun.

As noted, galaxies began as vast gravitational condensations of gas clouds that were created about 100 million years after the initial explosion, because of the random fluctuations of density within the primordial gas. If the mass of gas in a single such fluctuation exceeded a certain critical value—first deduced by the English physicist James Jeans (1877–1946)—gravitational contraction produced a stable cloud that did not disintegrate; and if enough such clouds collided, they coalesced gravitationally to form a galaxy. Initially, then, a galaxy was a fairly homogeneous gaseous structure with none of the gas in the form of stars or even showing any indications of star formation. To determine how stars finally emerged from such a smooth, undifferentiated medium, one must again consider the fragmentation of a uniform gas but this time take into account turbulences that arise when a fluid is in motion. Everyone has observed the variety of whirls and eddies that accompany a rapidly moving stream of water and perhaps has wondered about these small-scale irregular motions that are superimposed on the overall flow of the stream. Taken as a group, such motions in a fluid are called turbulences, and in recent years the subject of turbulences has evoked a great deal of interest in astronomy, physics, and engineering. This subject is of extreme importance in fluid dynamics because the motions of boats and planes are greatly affected by turbulences created in water and in air. It is very important in astronomy because it is believed that stars and planets originated from the turbulences that occurred in the gas and dust in galaxies shortly after the galaxies were formed.

It has been known for a long time that a smoothly flowing stream of liquid or gas will break up into many different irregular motions if the speed of the stream exceeds a certain critical value that is determined by the width of the stream, the density of the fluid, and the viscosity of the fluid. One can see intuitively that the tendency of a very viscous fluid to become turbulent is small because different parts of a viscous fluid tend to stick together, and hence to move together, so that the chance for differential motion and turbulence is quite small unless the fluid is moving rapidly. On the other hand, if the stream is very wide and the fluid quite dense, the chance for turbulence is great even for small streaming speeds. In a dense fluid, inertia is large, so that different parts of the fluid lag in their response to the overall flow, and differential motion occurs. If the stream is wide, different parts of the stream at large enough distances from the center fall out of step with the flow at the center and turbulence occurs.

All this can be expressed algebraically by introducing a dimensionless quantity called the Reynolds number, after the British physicist Osbourne Reynolds (1842–1912), who studied the flow of liquids in tubes of different diameters. He discovered that a smoothly flowing stream of a liquid or a gas will break up into turbulences of various sizes if the speed of the stream is so large that the number one gets when one multiplies the speed by the product of the width of the stream and the density of the fluid (grams per cubic centimeter) and then divides by the viscosity of the fluid is larger than 1,000. In other words, turbulence occurs in a fluid stream when the quantity

$$\frac{\text{speed} \times \text{width} \times \text{density}}{\text{viscosity}}$$

is larger than 1,000. And so, if the speed of any fluid is large enough, its motion is bound to break up into an irregular pattern of whirls and eddies of various sizes.

Consider now the gas swirling around the core in a rotating protogalaxy, the galaxy before the formation of stars. Apply to it the Reynolds criterion for turbulence. Since the viscosity of a gas equals the mean random thermal speed of the molecules multiplied by the product of the density of the gas and the mean free path of the molecules (the mean distance molecules move before they collide), the Reynolds number can be written

$$\frac{\text{speed of gas}}{\text{speed of molecules}} \times \frac{\text{width of stream}}{\text{mean free path of molecules}}.$$

In this form it is fairly easy to calculate the Reynolds number for the gas in a protogalaxy when the universe was about 2.5 billion years old. Since the speed of the molecules in a gas is determined by the absolute temperature of the gas, the molecules in the protogalaxy at that time could not have been moving much faster than one kilometer per second because the temperature of the universe had then dropped down to about 150°K, which is very cold, indeed. On the other hand, the speed of the rotating gas in the outer regions of the protogalaxy was of the order of 30 kilometers per second, as indicated by the observed rotational speeds of the spiral arms in known galaxies. The first factor in the above expression for the Reynolds number is thus about 30. Since the width of the stream of gas in a protogalaxy was about equal to the radius of the galaxy (some thousands of light-years) and the mean free path of the molecules was no more than a hundredth of a light-year, the second factor is very large. Thus, the Reynolds number was far greater than 1,000 in rotating protogalaxies, and so, turbulence necessarily occurred and the motion of the gas broke up into whirls and eddies of all sizes. This was the first in a series of steps that ended with the formation of stars, for, as will be shown, the turbulent motion produced local condensations that finally contracted into stars gravitationally.

Note that stellar formation began in the outermost regions of the protogalaxies because the differential stream velocities were greatest there, so that the turbulences there were very pronounced. Since this turbulent motion resulted in eddies of various sizes, it is important to determine whether eddies large enough to contract into stars were formed. Since billions of eddies were probably formed during this turbulent fragmentation of the primordial gas in the protogalaxies, there must have been a wide range of sizes among these eddies, and it is important to know how these sizes were distributed among the various eddies that were formed. If the predominant number was associated with eddies whose sizes were large enough to lead to star formation, stars must have formed quite quickly and the kinds of galactic structures that now exist must have appeared at a relatively early stage in the history of galaxies. Turbulence theory shows that the amount of translational kinetic energy per gram—measured by its translational speed—that an eddy has when it is formed depends on its size. The larger the eddy is, the faster it travels. From this it follows that very large eddies can be formed in appreciable numbers only if a great deal of kinetic energy is present in the streaming gas. Hence, there were probably very few large eddies present in the initial turbulence of the protogalaxies. On the other hand, although enormously large numbers of very small eddies were formed, the total amount of energy associated with these was very small because each such eddy had only a tiny amount of kinetic energy. From these two statements it is clear that most of the kinetic energy of the streaming gas was distributed among medium-sized eddies, which led to star formation.

When a hierarchy of eddies is formed in the turbulent motion of a gas, any particular single eddy has a very short life; it moves about for a distance more or less equal to its own diameter and then breaks up into smaller eddies. There is thus a continual transfer of kinetic energy from larger eddies to smaller ones until the energy is dissipated as heat. If there had been nothing more than this in the turbulent motion of the gas in a protogalaxy, stars would never have formed, but the force of gravity was also present and this led to gravitational contractions of colliding eddies where they were large enough. As the hierarchy of eddies moved about and collided with each other, they created a hierarchy of local condensations of various sizes. The small ones expanded again and dissolved very quickly, but those that had enough mass to contract gravitationally survived and grew by attracting more and more of the surrounding gas. This went on until these massive condensations had contracted gravitationally to such an extent that they were considerably denser than the surrounding gas and could no longer be supported by the hydrostatic gas pressure against the gravitational pull of the entire protogalaxy toward the center. This may be compared to the formation of raindrops or hailstones in the earth's

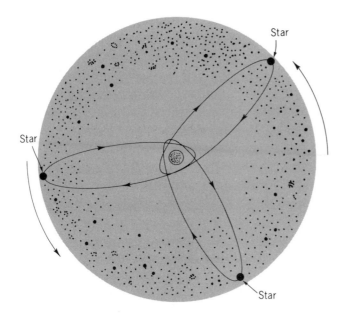

Stars condense from a hierarchy of turbulences (eddies) at the edge of a galaxy and fall toward the center.

atmosphere. Molecules of water vapor floating around high in the atmosphere cannot fall precipitously to the ground, because of atmospheric pressure. But if enough of these molecules come together to form condensations that are too dense to be supported by atmospheric pressure, a hailstorm or a rainstorm results.

Once a star was formed in the outer regions of a protogalaxy, it fell toward the center with ever-increasing speed, gathering more and more matter as it fell. It did not stop at the center of the protogalaxy but, owing to its inertia, passed right through and moved out again, although not quite as far out as the region in which it was born. The friction between it and the surrounding gas had robbed it of some of its kinetic energy, so that it stopped short of its starting point. It then fell back in again toward the center, and so on. These oscillations in a series of elongated elliptical orbits around the center continued until the star had lost much of its kinetic energy; it then settled down to a stable orbit near the center of the protogalaxy. One must now picture this as happening with billions of condensations so that there was literally a storm of stars falling into the center of the protogalaxy. As this star-formation process went on, the gaseous material in the protogalaxy continued to be used up until very little was left. The actual duration of this phase of star formation was probably no more than a few million years because the gravitational collapse of a condensation proceeds quite rapidly once enough material is present, and all such condensations must have occurred at about the same time.

When this process ended, the protogalaxy was transformed into an elliptical galaxy with very little gas and dust in it and with the stars forming a fairly symmetrical ellipsoidal distribution similar to that seen today.

The filamentary nebula in Cygnus, NGC6960; it consists of gas and dust.

The whirlpool nebula in Canes Venatici: spiral nebula NGC5194 (M51) and satellite nebula NGC5195. The magnificent spiral is dominated by dust lanes, the two most opaque of which lie on the inside of the two brightest arms. These arms and dust lanes describe almost perfect spirals as they emerge from the nucleus and unwind. NGC5195 is an irregular type, and its gravitational action probably accounts for the distortion of one side of NGC5194.

The stars and gaseous nebulosity in the region of Orion, showing some of the naked-eye stars.

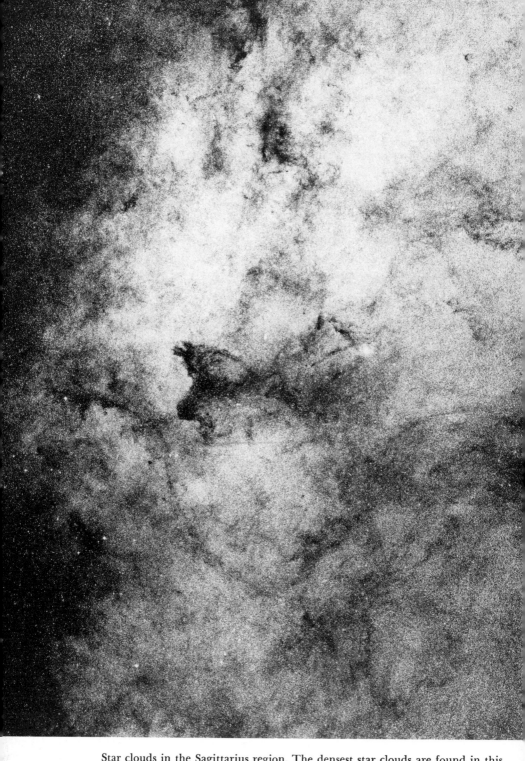

Star clouds in the Sagittarius region. The densest star clouds are found in this region of the Milky Way.

The Great Nebula in Andromeda, NGC224 (M31), and satellite nebulas NGC205 and NGC221. The spiral galaxy in Andromeda, at a distance of about 2 million light-years from the earth, is twice as large as the Milky Way and contains about twice as many stars.

Spiral nebula in Coma Berenices, NGC4565, seen edge on.

Cluster of galaxies in Hercules, 330 million light-years distant from the earth.
This cluster is extremely rich in spiral, elliptical, and barred galaxies.

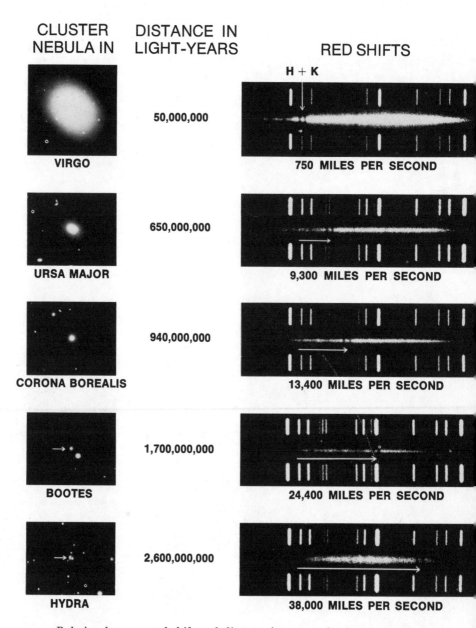

CLUSTER NEBULA IN	DISTANCE IN LIGHT-YEARS	RED SHIFTS
VIRGO	50,000,000	H + K 750 MILES PER SECOND
URSA MAJOR	650,000,000	9,300 MILES PER SECOND
CORONA BOREALIS	940,000,000	13,400 MILES PER SECOND
BOOTES	1,700,000,000	24,400 MILES PER SECOND
HYDRA	2,600,000,000	38,000 MILES PER SECOND

Relation between red shift and distance for extragalactic nebulas. Red shifts are expressed as velocities. The arrows indicate shift for calcium lines H and K.

The 200-inch telescope at Palomar, pointing north.

NGC1201

NGC 2811

NGC488

NGC2841

NGC3031　M81

NGC628　M74

Types of extragalactic nebulas. NGC1201 (*facing page, top*) is an elliptical that exhibits no spiral structure.

Nebula in Scutum Sobieski, NGC6611 (M16). This is a dusty region of the Milky Way where new stars are being born.

The Crab Nebula in Taurus, NGC1952 (M1). Recorded in 1054 by Chinese astronomers, this is the most famous of all supernovas. Its luminosity (some 10,000 times that of the sun) is supplied by the rotational energy of its central star, which is spinning 33 times a second and is a pulsar or a neutron star.

Globular star cluster in Canes Venatici, NGC5272 (M3). It contains only population-II stars. There is no dust or gas present.

Star cluster, open type, in Cancer, NGC2682 (M67). The oldest known cluster in the Galaxy, it consists of population-I stars.

The center of the spiral galaxy in Andromeda, showing the great concentration of population-II stars.

In the center of Andromeda, as in the center of our Galaxy, the red and yellow stars are population II—the oldest stars. The stars in the spiral arms of Andromeda belong to the much younger population I.

Mars *(top left)*, Jupiter *(top right)*, Saturn *(bottom left)*, and Pluto *(bottom right)*. The first three were photographed with the 100-inch telescope, Pluto with the 200-inch telescope.

The absorption lines (Fraunhofer lines) in a small portion of the solar spectrum. Each dark line arises from the interaction of some atom in the sun's atmosphere with the solar radiation streaming through the atmosphere. Most of the lines· are those of iron and other metals.

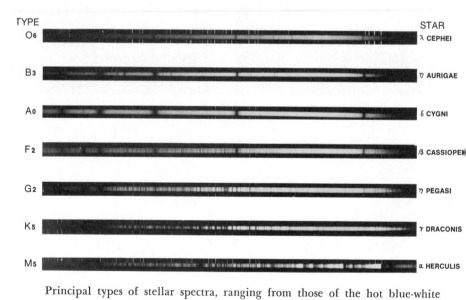

TYPE		STAR
O6		λ CEPHEI
B3		η AURIGAE
A0		δ CYGNI
F2		β CASSIOPEI
G2		η PEGASI
K5		γ DRACONIS
M5		α HERCULIS

Principal types of stellar spectra, ranging from those of the hot blue-white stars, types O and B, to the cool red stars, type M.

The region of the crater Clavius on the moon, photographed with the 200-inch telescope.

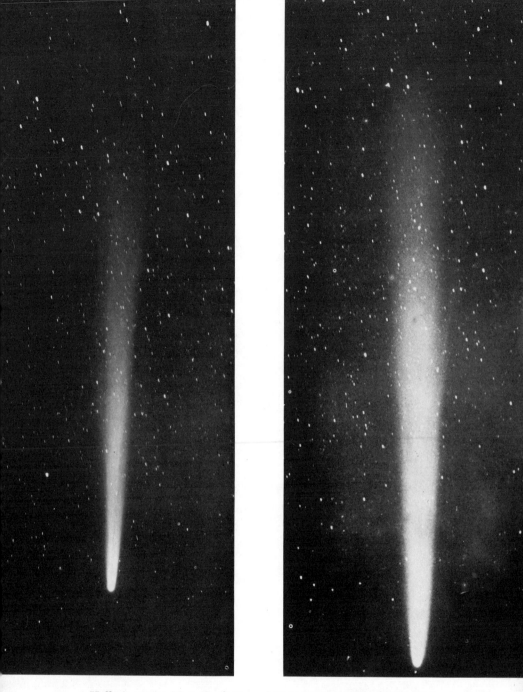

Halley's comet, May 12 (*left*) and May 15, 1910. The tail is 30 degrees and 40 degrees long, respectively. Photographed from Honolulu.

Planetary nebula in Aquarius, NGC7293, the end stage in the evolution of a star like the sun. The vast cloud of gas receding from the hot central star probably consists of iron atoms. The central star will probably become a white dwarf.

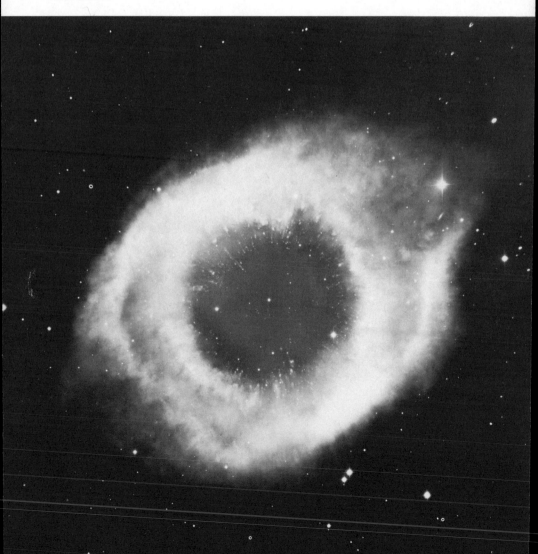

Galaxy in Centaurus, NGC5128, an unusual type. An explosion is occurring in the core of this galaxy. If such an explosion occurred at the center of our Galaxy, the energy emitted would be strong enough to alter life on the earth.

5 *The Nature of the Stars*

The Two Stellar Populations

The universe contains two distinct star populations, which astronomers have labeled population I and population II. These designations may seem a bit puzzling to some, because the population-II stars are the older stars and hence those that were born first, whereas the population-I stars, among which are stars like the sun, came on the scene some billions of years later. One would have expected "population I" to refer to the stars that appeared first in the universe and "population II" to those that came later instead of the other way around.

If the division of stars into two populations had been introduced after stellar ages and stellar evolution were known, the numbering would probably now be the reverse of what it is, but when the German-American astronomer Walter Baade (1892–1960) discovered in the early 1940s that there are two distinct populations of stars, hardly anything was known about the aging of stars; he assigned the numbers I and II the way he did because he wanted the sun to be designated as a population-I star. It was not until some 20 years later that it was discovered that the sun is a relatively young, second-generation star, whereas the stars Baade called II are very old, first-generation stars. Baade's discovery of the two stellar populations stemmed from his work on Andromeda in 1943 when he succeeded in resolving into stars, on a photographic plate, the central region of Andromeda and its two companion galaxies NGC205 and NGC221. Careful comparison of the stellar composition of the spiral arms of Andromeda with that of its core revealed that spiral galaxies generally consist of two distinct stellar populations, one associated with the spiral arms of the galaxy (population I) and the other with the core (population II).

Until Baade did his pioneering photographic work on Andromeda and its companions, the existence of individual stars in the cores of galaxies had never been demonstrated observationally, although it had been accepted as a fact. The reason that individual stars had not appeared on photographs of galactic cores earlier is that such photographs were taken with ordinary photographic plates, which are sensitive to blue, but not red, light. On such plates the core of Andromeda shows up as a very bright smooth object, with no indication that it consists of individual stars, whereas the individual stars in the spiral arms can be distinctly seen, intermixed with gas and dust. Baade was greatly impressed by the fact that the spiral arms of Andromeda are marked by the coexistence of very luminous blue-white stars (blue-white giants) and gas and dust clouds, whereas the core shows no dust, gas, or stars on blue-sensitive photographic plates. Baade reasoned, quite correctly, that if one used red-sensitive photographic plates for the core, the individual core stars would show up. The redness of galactic cores is a marked characteristic of galaxies. The color difference between the arms of Andromeda and its core is so pronounced that Baade once mistook the core of Andromeda, which he had focused on visually, for a forest fire. If the color is caused by red-colored stars, special red-sensitive plates must be used to record individual images of these stars. Moreover, such plates can also reveal the regions of ionized hydrogen, called H_{II} regions, along the spiral arms. Since hydrogen is the most abundant element in the universe, one would expect to detect great quantities of it along the spiral arms under appropriate circumstances, but Hubble's exhaustive photographic survey with blue-sensitive plates of one of the arms of Andromeda revealed no hydrogen at all. The reason for this is that the clouds of ionized hydrogen emit great quantities of red light when the ionized hydrogen captures electrons.

Taking all this into account, Baade decided to study Andromeda with red-sensitive plates, and when he did so, everything fell neatly into place. The individual red stars in the core, as well as those in the two companion galaxies, appeared as distinct images; and the H_{II} regions were "strung out like pearls along the spiral arms." The red-sensitive plates brought out the H_{II} regions by the hundreds and showed them, together with their blue and blue-white giant stars, to be deeply imbedded in the dust of the spiral arms. Baade designated the stars in the spiral arms of the galaxies as population I because they are the stars that were first studied and understood by astronomers. He pointed out that the gas and dust are the primary constituents of the spiral arms, whereas the population-I stars associated with the gas and dust are secondary phenomena, with such stars being formed from the dust and gas right up to the present day. One must, indeed, consider the dust and gas clouds, like those in the Orion nebula, as the "breeding place of young population-I stars." Baade designated the stars in the cores of galaxies and in the globular clusters as population-II stars because this population was discovered to be a distinct new class of stars, different in their overall characteristics from the population-I stars found in the arms of the spiral galaxies and in the irregular galaxies. Baade's great contribution was to show that globular clusters, elliptical galaxies, and the cores of spirals consist of pure population-II stars. Note that the designations population-I and population-II were not introduced to indicate the ages of these two classes of stars.

There follows now a very remarkable conclusion about the molecules and atoms that constitute all bodies—sun, planets, comets, living organisms, and so on—in the solar system: they all must have been part of some star billions of years ago. How striking that, as long ago as 60 B.C., the Latin poet Lucretius surmised this when he stated that "this snow-flake was a flame; the flame was the fragment of a star." Lucretius had only his intuition to guide him to such a conclusion, but today we have direct proof, in the existence of the two stellar populations, that every atom in every human body existed, in a much simpler form, in the deep interior of some long-dead population-II star. Spectral analysis of the atmosphere of population-II stars confirms that they were born from the primordial gas, which consisted of hydrogen and helium only; hardly any of the spectral lines of the metals or other heavy elements are found in the spectra of these stars. This means that life, and organic elements in general, could not have evolved on planets revolving around population-II stars, because the various heavy elements—carbon, phosphorus, sulfur, iron, among others—that are essential for life, and large organic molecules in general, were not present when population-II stars and their planets were formed. On the other hand, the population-I stars and their planets were, and still are being, formed from dust and gas that consist of the heavy elements as well as of hydrogen and helium. Where did this gas and dust mixture of hydrogen, helium, and the heavy elements come from? From the deep interiors of population-II stars, where at a certain stage in the lives of these stars the temperature and pressure were so high that the heavy elements were synthesized from hydrogen and helium. This took billions of years, after which vast stellar explosions ejected great quantities of hydrogen, helium, and the heavy elements into space. It was from this gaseous mixture enriched with heavy elements that dust clouds and, later, population-I stars were formed; in this process planets that can sustain life, like the earth, were also formed.

Vital Statistics of Stars

All stars, regardless of the population to which they belong, have in common certain characteristics that mark them as stars. Speaking quite broadly, stars are gaseous, self-luminous, spherical bodies having the same general characteristics as the sun. Since almost all of the information that can be obtained about a star comes to earth in the form of electromagnetic radiation, one must observe as much such radiation as possible, using both optical and radio telescopes. Stars, in general, emit relatively small quantities of radio waves as compared to their optical emissions. The discussion here is therefore limited to what can be learned from the visible electromagnetic radiation (that is, light) emitted by stars.

STELLAR LUMINOSITIES

The total amount of radiation emitted per second in all directions by a star is called its luminosity, which may be expressed in terms of the known solar luminosity. Thus, a luminosity of 10 means the star is emitting 10 times as much radiant energy per second as the sun is. But to find the luminosity of a star, its distance must be known. If one knows how far away a star is and how bright it appears, one can calculate its luminosity. The reason for this is that a simple algebraic formula relates the luminosity of an object to its distance and its apparent brightness. If the distance of a source of light is increased, its apparent brightness diminishes by the square of the distance. Thus, if the sun were 2, 3, or 4 times as far away as it now is, its apparent brightness would decrease by a factor of 4, 9, or 16, respectively. Measuring the distance of a star is, then, very important.

If the position of a star relative to other stars in the same general part of the sky is carefully determined at two different times of the year six months apart, it will be found to have changed. One finds, in fact, that the stars in general appear to shift their positions back and forth every six months. This apparent shift, which is caused by the annual motion of the earth around the sun, is related, by means of a simple algebraic formula, to the diameter of the earth's orbit around the sun and to the distance of the star. In fact, one finds that

$$\text{star's distance} = \frac{\text{diameter of earth's orbit}}{\text{semiannual shift of the star}},$$

which simply says that the closer the star is, the larger is its apparent shift. This formula has been used extensively to determine the distances of more than 10,000 nearby stars, but there is a limitation to its use, for the apparent semiannual displacements of very distant stars, which are caused by the earth's motion, are too small to be measured. In fact, the above formula is not reliable for stars more than a few hundred light-years away. Most of the stars that can be seen with the naked eye are within 1,000 light-years, with the closest being about 4.5 light-years away. The stars that crowd together to form the Milky Way are thousands of light-years away.

Since the sun is used as a standard for stellar luminosities, its luminosity is given in traditional energy units—that is, calories or ergs. The calorie is the amount of energy required to increase the temperature of one gram of water by one degree centigrade; 80 calories are needed to melt one gram of ice. The erg, which is the standard unit of energy used in physics and astronomy, is much smaller than the calorie; in fact, 1 calorie equals about 42 million ergs.

The amount of radiant energy emitted by the sun in all directions is such that, if there were no atmosphere to absorb the solar radiation, each square centimeter of the earth's surface would receive 2 calories every minute. From this number, which is called the solar constant, and from the distance of the sun, one can calculate the solar luminosity, which is written as 4×10^{33} ergs per second—that is, 4 billion trillion trillion ergs per second. One may put this differently and perhaps more dramatically by saying that the sun radiates in one second enough energy to melt a column of ice

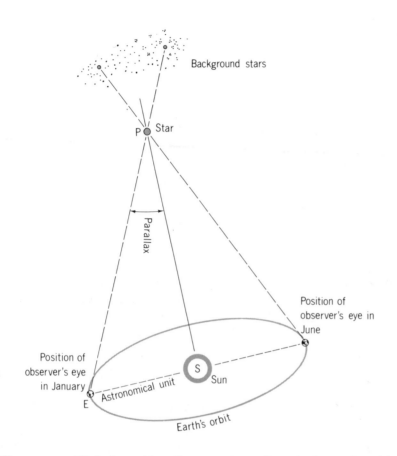

The apparent shift in the position of a star against a distant background resulting from the earth's motion. This shift is called the parallax of the star. From it and the knowledge of the earth's distance from the sun (the astronomical unit) the star's distance can be determined.

3.5 miles in diameter and extending from the earth to the sun (93 million miles). Another way to express this is to use the Einstein relationship between energy and mass, $E = mc^2$. This equation tells us that if a given amount of mass m is transformed into an equivalent amount of energy E, the amount of energy released equals the mass m multiplied by the square of the speed of light, c^2. If m is expressed in grams and c in centimeters per second (30 billion centimeters per second), then E is obtained in ergs. Thus, one gram of matter is equivalent to 900 million trillion ergs. This is enough energy to lift about one million tons to a height of 6 miles above the surface of the earth. If one uses the Einstein energy-mass relationship, one finds that the energy emitted by the sun every second is equivalent to 4.5 million tons. This means that the sun is transforming 4.5 million tons of mass into energy every second to obtain the energy it needs to keep it as luminous as it is.

Using the measured distances of stars and their apparent brightnesses, one finds an enormous range of values for the stellar luminosities. The most luminous stars are about 70,000 times as luminous as the sun, whereas the sun is thousands of times as luminous as the faintest known stars. A few examples, using well-known stars, will illustrate this important point. One of the most luminous visible stars is the blue-white giant Rigel, in Orion; it is about 60,000 times as luminous as the sun. Betelgeuse, which is a red supergiant in Orion, is about 12,000 times as luminous as the sun, whereas the red supergiant Antares, in the summer constellation Scorpio, is about 8,000 times as luminous as the sun. The beautiful white star Sirius, which rises on the heels of Orion in the winter months and is about 9 light-years away, is about 20 times as luminous as the sun, whereas the white star Vega, which rides high in the evening summer sky, is about 30 times as luminous. An example of an intrinsically very faint star is Proxima Centauri, which revolves around Alpha Centauri; the sun is about 10,000 times as luminous as Proxima. There are, all told, about 100 stars within a distance of 10 light-years of the sun that are anywhere from a few hundred to a few thousand times less luminous than the sun.

This wide range of luminosities tells us something of great importance about the rate at which various stars age. Since the luminosity of a star is determined by the rate at which it is burning its nuclear fuel, it may be concluded that the very luminous blue-white stars like Rigel are living very riotously and consuming their fuel at such a prodigious rate that they cannot continue doing so for very much longer. The amount of nuclear fuel—hydrogen and helium—that stars are born with is about the same whether the star is one like Rigel or one like the sun. If, then, Rigel is consuming its fuel some 60,000 times faster than the sun is, it will exhaust its fuel and therefore age that much faster than the sun does. Thus, given their great luminosity, Rigel and stars like it must be very young, for they could not have sustained this vast outpouring of energy for more than a few million years in the past. The very faint red stars, or red dwarfs, are aging and evolving, and hence changing, very slowly. Since the faint red stars change very slowly, one cannot say just how old they are; some may be quite young, but most of them are probably very old. A star like the sun ages at a rate that is somewhere between that of the red dwarfs and the superluminous blue giants.

The red supergiants like Betelgeuse and Antares, which are some 10,000 times as luminous as the sun, are also using up their nuclear fuel very rapidly and therefore evolving very fast, but they are not young stars. They are, indeed, quite far along in their evolution and probably began their lives as blue-white giants about 100 million years ago. The red supergiants are near the end of their lives and have reached a stage where they have exhausted all their hydrogen and are rapidly using up their helium and other nuclear fuel.

A question that naturally arises here is what physical factor determines whether a star at birth will be a blue-white one, destined to live a short but very feverish life, or one that ages much more slowly and gracefully, like the sun. The mass of the star at its birth is the determining factor here; the more massive a star is, the more rapidly it ages.

COLORS OF STARS

A casual glance at the night sky leaves one with the impression that all the stars are more or less white, but a more careful examination of individual stars reveals that they vary considerably in color. This has led astronomers to the classification of stars by color, which ranges from red to blue-white. These color designations, however, are only relative ones, and the reader is cautioned not to fall into the error of thinking that a red star emits only red light. In fact, stars, regardless of the color class to which they are assigned, emit light of all colors but not with equal intensities. If a star emits much more red light than it does green, blue, or violet light, it looks red and is called a red star. On the other hand, if the most intense radiation from the surface of a star is in the yellow regions of the spectrum, as in the case of the sun, the star looks yellowish. In all of these cases the green, blue, and violet colors are also present but in relatively insignificant quantities. If the radiation from a star is relatively rich in the green, blue, and violet colors, the star looks white or distinctly blue-white, even though the reds and yellows may be quite intense. Indeed, the blue-white star Rigel emits more red and yellow light per second than the faint red dwarfs do. Sirius and Vega are good examples of white stars, and Arcturus is an excellent example of an orange star.

TEMPERATURES OF STARS

The color of a star is of special interest because, from it, one can determine the surface temperature of the star. The color of a glowing coal or a furnace depends on its temperature; the higher its temperature, the whiter it is. When the temperature of a furnace is below 1,000°, its hot interior has a deep reddish glow; but as the temperature is increased, its color changes from red to yellowish and finally to a dazzling white as the temperature approaches 10,000°. This dependence of color on temperature was studied intensively by physicists near the end of the nineteenth century. It aroused considerable interest because the experimental results disagreed violently with the results deduced from the classical theory of radiation, Maxwell's electromagnetic theory of light, which pictures radiation as an electromagnetic wave. Since the intensities of the various colors emitted by a hot body at a given temperature as predicted by this wave theory disagreed with the observed intensities emitted by the body, it was necessary to introduce a change in the wave theory to bring it into agreement with the observations. The great German physicist Max Planck (1858–1947) was the first to understand what was needed, and in the year 1900 he proposed the revolutionary idea that a hot body emits radiant energy in the form of discrete pellets, or quanta, instead of in the form of continuous waves. This was the genesis of the quantum theory, which led to the concept of the photon, a particle of radiation, as developed by Einstein in a famous paper a few years later. Using his quantum hypothesis, Planck derived a mathematical formula for the radiation emitted by a hot body that agrees perfectly with the observed radiation for a given temperature. Since this formula relates the intensities of the various colors emitted by the hot body to its temperature, one can use this formula to determine the temperature of the hot body from the observed intensities of the emitted colors. Using this procedure to determine the surface temperatures of the stars, scientists have found that these temperatures range from about 3,000°K for red stars to about 40,000°K for the hottest, and therefore bluest, of the blue-white stars. The surface temperature of the sun and of yellow stars like it is about 5,600°K, whereas that of a white star like Sirius is about 10,000°K; the surface temperature of Rigel is about 15,000°K. The reader must keep in mind here that these temperatures refer only to the surface layers of the stars and not to their interiors, where the temperatures are higher by at least a factor of 100.

STELLAR SPECTRA

The radiation emitted by a star is a mixture of various colors of differing intensities, which was first discovered when

Newton passed light from the sun through a prism and observed that white light was spread out into an array of colors that ranged continuously from red to violet; this array of colors is called a spectrum. When Newton observed the solar spectrum, he failed to detect in it innumerable fine dark lines that cross each color. These lines can be seen if the light from the sun passes through a narrow slit before it strikes the prism; they are then seen to be parallel to the slit, as was first observed in 1802 by the British scientist William Hyde Wollaston (1766–1828). In 1815 the German physicist Joseph von Fraunhofer (1787–1826) obtained a solar spectrum in which hundreds of these dark lines—now called Fraunhofer, or absorption, lines—were clearly shown.

With the development of atomic theory and an understanding of the way atoms emit and absorb electromagnetic radiation, it became clear that the Fraunhofer lines in the spectrum of the sun or any other star stem from the interaction of the atoms in the star's atmosphere with the electromagnetic radiation coming from the deep interior. Since an atom consists of a positively charged nucleus with negatively charged electrons circling it in various orbits, it is clear that these electrons will be affected by a stream of electromagnetic radiation moving through the atom. An electron in any given atom can absorb some of this radiation and use this absorbed energy to move into another orbit. If one pictures the electrons in the many different atoms of the various elements in the atmosphere of a star absorbing radiation in this way, one can see why absorption lines occur in stellar spectra. The reason the absorption shows up as discrete lines in a spectrum is that the electrons inside atoms move in discrete orbits and the jumping of an electron from one orbit to a larger orbit when it absorbs radiation gives rise to a single definite absorption line. Since the electrons in each atom of a particular kind have a definite set of orbits, all the atoms of a given element have the same absorption lines, which are different from those of the atoms of other elements.

Many years of research and analysis have led to a precise identification and classification of the spectral lines of the various elements, so that the spectrum of a substance can now be used to find its chemical composition. The absorption lines in the spectrum of a star thus permit one to identify the various elements in the star's atmosphere. In this way, most of the elements found on the earth have been found in the atmospheres of stars.

Astronomers have found that the absorption lines in the spectra of stars can be used to classify stars into spectral groups that correspond to stellar colors and surface temperatures. This classification is based on the observation that the spectral lines of the various elements are not present in equal intensities in the spectra of all stars. These differences in the spectra of stars were first interpreted as arising from the differences in the chemical composition of the atmospheres of these various stars, but this interpretation was shown to be untenable in general. It is known that the atmospheres of population-II stars have a different chemical composition from those of population-I stars, but the spectral differences referred to above exist between the spectra of stars that belong to the same population. Thus, one finds variations in the intensities of spectral lines among stars that form a single family, stars that belong to the same cluster and hence were born from the same chemical mixture. This led astronomers to look for another cause for these spectral differences between stars of the same chemical composition, and they found it when they discovered that the spectral differences could be correlated to color differences.

If one arranges the spectra of stars in a continuous vertical array with the spectra of the bluest stars on top and those of the reddest on the bottom, one observes a gradual change in the appearance of the spectra and in the intensities of lines. At the very top, in the spectra of the very bluest stars, the spectral lines of ionized helium are most intense, but they diminish in intensity as one moves down while the intensity of the lines of neutral helium steadily increases. These reach a maximum for stars like Rigel and then steadily decrease as the lines of hydrogen become more and more prominent, becoming most intense for the white stars like Sirius. As one moves down the sequence the hydrogen lines begin to fade and the spectral lines of the neutral and ionized metals, such as iron and calcium, increase in intensity, becoming most pronounced in yellow stars like the sun and in orange stars like Arcturus. As one approaches the spectra of the red stars, the lines of the metals fade away and the closely packed lines—so closely packed that they merge into bands—of such strongly bound molecules as titanium oxide and hydrogen oxide appear.

How is one to interpret the variations in this sequence of spectra? One can find the answer to this question in the correlation that exists between these variations and the colors of the stars. Since the colors of the stars are related to their surface temperatures, the differences in the spectra of stars of different colors are surmised to be caused by differences in their surface temperatures. Although the observational evidence alone gives strong support to this conclusion, additional, and conclusive, support for it is found in atomic theory. The electrons and the nuclei of atoms, like the atoms within molecules, are held together by their mutual electromagnetic interactions, but the strength with which electrons are bound to nuclei inside atoms is not the same for all kinds of atoms, and the strength with which atoms are bound together in molecules is not the same for all kinds of molecules. This binding strength of electrons in atoms and atoms in molecules is measured by the energy required to tear an electron out of an atom or to tear a molecule apart. Since electrons are electrically charged particles and light is an electromagnetic disturbance, they can be torn away from nuclei of atoms by irradiating the atoms with the proper kind of light. Molecules can be disrupted in the same way because the atoms in molecules also respond to the electromagnetic vibrations of light.

Consider now a hierarchy of molecules and atomic structures arranged schematically on a ladder, with those in which the constituent particles are most weakly bound placed at the bottom of the ladder and those with very strongly bound constituent particles at the top. At the very bottom then are the very weakly constructed large organic molecules like those that constitute living organisms, and at the very top are the ionized atoms such as ionized helium, carbon, and oxygen. Molecules such as titanium oxide and hydrogen oxide lie pretty far up on this ladder, just below the metals, and hydrogen lies below helium, which in turn lies below ionized helium. If such atoms and molecules form a gaseous mixture through which thermal radiation is streaming, the radiation affects, and in turn is affected by, this gaseous mixture. If the radiation is cool, it has only a small effect on the gas particles, causing them to rotate in various ways and to move about at fairly low speeds. But things change drastically as the temperature of the radiation rises to 1,000°K or more; most of the photons in this radiation are now so energetic that they cause the gas particles to move about much more rapidly when they collide with these gas molecules. Moreover, these photons are now energetic enough to set the molecules and atoms vibrating in highly excited states upon being absorbed by the electrons in the atoms and molecules. Both of these phenomena will tend to disrupt the molecules and ionize the atoms. At about 1,000°K most organic molecules are moving about so fast that they break up when they collide or they are broken up by the energetic photons, but most atoms are not changed; they are only speeded up because 1,000°K radiation is still too cool to do more than this. At 3,000°K, only the most tightly bound diatomic molecules, such as some of the oxides, survive, but only those atoms with weakly bound electrons are excited by the photons, which even at 3,000°K are not very energetic. As the temperature of the radiation rises, all molecules are disrupted and some of the tightly bound atoms are excited and even ionized. Finally, at a high enough temperature most of the photons in the radiation streaming through the gas are so energetic that only the most tightly bound neutral and ionized atoms, such as helium, can survive.

If one applies this analysis to the atmosphere of stars, one can explain most of the characteristic features of the spectra of these stars. Since the red stars are the coolest and the blue-white are the hottest, the spectra of the former are dominated by the molecular bands of such molecules as titanium oxide and hydrogen oxide and the spectra of the latter are dominated by the lines of neutral and ionized helium. Although this general description of how the surface temperature of a star, and hence the temperature of the radiation it emits, determines the overall spectral characteristics of the star's atmosphere is correct in a general way, it does not give us any of the details of the atomic processes that are involved in the formation of a star's spectrum. Modern atomic theory, which had its birth in the theory of the atom proposed by the Danish physicist Niels Bohr (1885–1962), and evolved into its present powerful form with the discovery of the wave properties of matter, has been applied extensively to the analysis of stellar spectra, and the spectral lines deduced from the theory are in complete agreement with the observed spectral lines of the various stars. Indeed, the theory of the formation of spectral lines is so accurate that one can deduce the surface temperatures of the stars as well as the concentrations of the various elements in the atmosphere from the intensities of the spectral lines.

Astronomers have found it convenient to divide the stars into seven broad spectral classes to which the letters O, B, A, F, G, K, and M have been assigned. The O-type stars are the hottest blue-white stars, with intense ionized and neutral helium lines, and the M-type stars are the cool red stars.

Yellow stars such as the sun are G-type stars, and the white stars Sirius and Vega are A-type stars.

STELLAR DIAMETERS

The diameters of the stars, which vary enormously as one goes from the intrinsically faint stars, the dwarfs, to the very luminous stars, the supergiants, are important stellar characteristics because a number of other characteristics—surface temperature, surface gravity, luminosity—are related to the diameters. The only star whose diameter can be measured directly is the sun because it is the only star whose apparent size—the angular diameter of its visible disk—can be measured. Since the disk of the sun subtends 0.5° at the eye of an observer on the earth and its distance is 93 million miles, one finds geometrically that its true diameter is approximately 1 million miles. One cannot compute the diameter of any of the other stars in this way, because they appear to the observer as mere points of light. Their true images are not revealed even by the largest telescopes now in use; if one had a 400-inch optical telescope, one could see the true images of the very largest stars, such as Betelgeuse and Antares, and measure their angular diameters. Stellar diameters must therefore be measured by an indirect procedure—the use of an algebraic formula that relates the luminosity of a star to its surface temperature and its radius. It can be shown, from the general properties of radiation, that the rate at which each unit of area of a star's surface radiates energy is proportional to the fourth power of its absolute temperature. It follows from this that the total luminosity of a star is proportional to the product of the fourth power of its surface temperature and its surface area. Since the surface area of a star is proportional to the square of its radius, the luminosity of a star is proportional to the product of the square of its radius and the fourth power of its surface temperature. This is written algebraically as follows: $L \propto R^2 T^4$, where R is the star's radius. Since the luminosity L of a star can be found from its distance and its absolute temperature T is known from its color, this relationship can be used to find the star's radius.

This can be illustrated with a few examples where the quantities R, T, and L are all taken relative to the corresponding solar quantities. Since the red dwarfs have surface temperatures about $\frac{1}{2}$ that of the sun, their luminosities would be $\frac{1}{16}$ of the sun's (the contribution of T^4 to L; the fourth power of $\frac{1}{2}$ is $\frac{1}{16}$) if the red dwarfs were as large as the sun. But the luminosities of red dwarfs are about $\frac{1}{1000}$ that of the sun; hence they must be smaller than the sun. In fact, R^2 for the red dwarfs must be about $\frac{1}{62}$ of what it is for the sun ($\frac{1}{16}$ times $\frac{1}{62}$ is about $\frac{1}{1000}$) and the red dwarfs' R is therefore about $\frac{1}{8}$ the sun's R. The surface temperature of the supergiant Betelgeuse is $\frac{1}{2}$ the sun's, but its luminosity is 10,000 times as great. From this it follows that Betelgeuse must be enormously larger than the sun. Since T^4 for Betelgeuse is $\frac{1}{16}$, R^2 must be 160,000 and R is 400, an incredible size. The volume of Betelgeuse is thus large enough to accommodate about 100 million suns; if it were placed at the center of the solar system, its surface would extend out to about Jupiter's orbit. The red supergiants, in general, have diameters that are more than 100 times the sun's diameter. What about the white stars like Sirius and Vega that are about 25 times more luminous than the sun? Since their surface temperatures, which are about twice the sun's, contribute a factor of 16 to their luminosities, their radii must be 25–30 percent larger than the sun's to give them their observed luminosities.

The blue-white giants, with surface temperatures about 3 to 4 times the sun's surface temperature, are about 50,000 times as luminous as the sun. Since the fourth power of the temperature contributes a factor of about 100 to this luminosity, the surface area must contribute a factor of about 500, so that the radii of these stars are 20–25 times the sun's radius. It

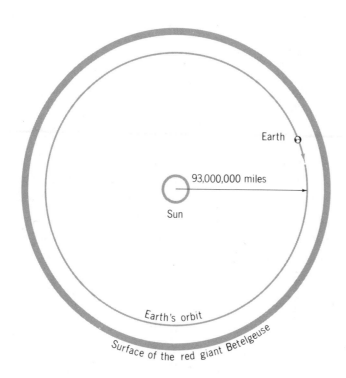

If the red supergiant Betelgeuse in Orion were placed at the center of the solar system, its outer surface would extend far beyond the earth's orbit.

is seen, then, leaving the red supergiants out of account for the moment, that the diameters of the stars range continuously from 0.1 to 25 times the sun's diameter as from the red dwarfs to the blue-white giants.

These stars, then, form a continuous sequence in which the change in their luminosities, colors, surface temperatures, spectral classes, and diameters are all correlated and change continuously as one moves from the cool red stars to the hot blue-white stars. Since the red supergiants do not fit into this sequence—their radii and luminosities are much too large— they must represent a completely different stage in a star's evolution from that represented by the stars in the sequence described above, which astronomers have labeled the main sequence because they are, by far, the most plentiful stars in earth's part of the Galaxy.

There are three other groups of stars that do not belong to the main sequence but whose characteristics are of great interest and importance in the evolution of stars. First is a group of cool yellow and red stars called giant stars, whose luminosities are all about the same and about 100 times the sun's luminosity and whose diameters are about 30 times the sun's diameter. Then there is a group of very faint hot, white stars that are called white dwarfs. Since the surface temperatures of these stars are about twice the sun's surface temperature but their luminosities are only one-thousandth of the sun's luminosity, their diameters must be about one-hundredth of the sun's diameter; in other words, the white dwarfs are no bigger than the earth. Finally, there are the recently discovered pulsars, which in general are too faint to be seen but which emit pulses of radio waves with great regularity. One pulsar has been photographed optically at the center of the Crab Nebula; indeed, it is itself the central star of that nebula and is the source of the energy that gives the nebula its great luminosity. Since the surface temperatures of pulsars are very high (at least 4 times the sun's surface temperature, as indicated by the very blue light emitted by the pulsar at the center of the Crab Nebula) but their optical luminosities are very small, the pulsars must be extremely small objects. In fact, they are 10 to 15 miles in diameter.

The question of how to account for such a wide range in stellar sizes remained unanswered until modern electronic computers enabled astrophysicists to trace the evolution of a star from its birth to its old age. It then became clear that the five different groups of stars described above represent different stages in the evolution of individual stars. The theory of stellar evolution shows that stars begin their lives as members of the main-sequence group but then, as they use up their nuclear fuel, change to giant stars, and so on. The determining factor in this evolutionary process is perhaps the most important of the physical stellar characteristics: the masses of stars.

STELLAR MASSES AND DENSITIES

Although the luminosities and diameters of stars vary widely from one group to another, the stellar masses do not; the mass of the gaseous cloud that ultimately becomes a star is an extremely critical characteristic that determines how hot and luminous the star will be at its birth and how rapidly it will age and evolve. The more massive stars are hotter, more luminous, and evolve faster than the less massive stars; even a small difference in mass means a large difference in the other stellar characteristics. The mass of a star can be measured directly only if its gravitational effect on some other body can be observed. One can then use Newton's law of gravity to deduce the mass of the star from this gravitational effect. Thus, the mass of the sun can be deduced from the earth's period of revolution and the earth's mean distance from the sun; Newton's law of gravity leads to the fact that the mass of the sun is proportional to the cube of the earth's mean distance from the sun, divided by the square of the earth's period of revolution. This same relationship holds for any of the other planets revolving around the sun, as was discovered empirically by Kepler some 20 years before Newton was born; thus, one may write, in general,

$$\text{mass of sun} \propto \frac{(\text{mean distance of planet from sun})^3}{(\text{period of revolution of planet})^2}.$$

Using the data for the earth, this formula gives the sun's mass as 2 billion trillion trillion grams (written as 2×10^{33} grams), or about 340,000 times the earth's mass.

This same formula can be used to find the mass of any star if it is a member of a binary system, a system of two stars that are bound to each other gravitationally and revolve around their common center of mass. If the true orbit of each star in the binary system around the common center of mass can be plotted or calculated, the mass of each star in the system can be computed from the observed period of revolution of the two stars and their mean separation. The procedure is somewhat more complicated than, but essentially the same as, that outlined above for finding the mass of the sun from the elements of a planet's orbit around it. Since there are many stellar binary systems in the Galaxy, many stellar masses have been determined directly from binary orbits; the surprising result is that the masses of stars in general do not differ much from that of the sun. The intrinsically very faint red dwarfs are about a fifth as massive as the sun whereas the luminous blue giants are about 25 times as massive. Stellar masses are found to have a 125-fold range, which is relatively small when compared to the 100 million-fold range of luminosities for the same stars. There is an important reason for this restriction on the masses of stars.

Since the determination of the masses of the two companions of a binary system is a tedious and long-drawn-out process, it is fortunate that a relationship between the luminosity and mass of a star exists that permits us to determine the masses of stars very quickly from their luminosities. This mass-luminosity relationship was discovered empirically when it was observed, from the calculated masses of binaries, that the more massive stars are also the more luminous ones. A statistical analysis of the empirical data shows that the luminosities of stars vary almost as the fourth power of their masses. We write this as $L \propto M^4$. This means that a star that is about 10 times as massive as the sun is about 10,000 times as luminous, whereas one that is 5 times less massive is about 700 times less luminous.

That a mathematical relationship between the luminosity and mass of a star must exist can be deduced from very general physical principles. The larger the mass of a star is, the larger must the internal pressure be, to prevent the star from collapsing under its own weight. But for intense gas pressure to build up inside a star, high internal temperatures are needed, and this means large luminosities because the internal temperature of a star determines the rate at which the star generates energy. Consequently, stars cannot be too massive, for if they were, the pressure of the energy generated in the star would disrupt it. The mass-luminosity relationship has enabled astronomers to determine the masses of hundreds of stars that do not belong to binary systems.

From the mass of a star and its size can be calculated its mean density—that is, the compactness of its mass—which is defined as the star's mass divided by its volume. Here an enormous range of values is found because although the masses of stars do not differ much, their volumes, from the smallest to the largest, differ by many trillions. To have a basis for comparison with the densities of ordinary kinds of matter here on the earth, note that the density of water is 1 gram per cubic centimeter and the density of platinum, one of the densest known metals, is about 22 grams per cubic centimeter. Stars like the sun, whose volumes are about one million times as much as that of the earth have mean densities of about 1.5 grams per cubic centimeter (not much denser than water), whereas the red supergiants like Betelgeuse have mean densities less than one-tenth of a millionth of a gram per cubic centimeter, which is about as dense as what is called a good vacuum here on the earth. Red and yellow giants have mean densities of about one-thousandth of a gram per cubic centimeter, and the blue giants like Rigel have densities of about one-hundredth of a gram per cubic centimeter. When one comes to the intrinsically faint stars like the red and white dwarfs, one runs into much larger densities. The red dwarfs have mean densities of about 25 grams per cubic centimeter and the white dwarfs have densities that range from 100,000 to some 10 million grams per cubic centimeter; a matchbox full of matter from a typical white dwarf would weigh a few tons on the earth. Until pulsars were discovered, the white dwarfs were the densest known objects in the universe, but that distinction now goes to the pulsars, whose densities are of the order of 1,000 trillion grams per cubic centimeter.

The Hertzsprung-Russell Diagram

A complete modern theory of the internal structure and evolution of stars recognizes broad correlations between the gross physical parameters of stars, but long before any such reliable theory was developed the Danish astronomer Ejnar Hertzsprung (1873–1967) and the American astronomer Henry Norris Russell (1877–1957), working independently, had discovered a very remarkable empirical correlation between the luminosities and surface temperatures of stars that proved to be extremely important for, and a great stimulus to, the theory of stellar evolution. Hertzsprung and Russell showed that the full significance of the luminosity-color relationship is revealed only when the stars are separated into such groups as main-sequence stars and giants. Indeed, the designations of stars as giants and dwarfs were first introduced by Hertzsprung in the early part of this century when he observed that stars of spectral type M, the red stars, can be divided into two distinct groups, one consisting entirely of intrinsically faint stars, with luminosities about one-thousandth of the sun's, and the other, of very luminous stars, a few hundred times more luminous than the sun. This discovery, which was later extended to orange and yellow stars by Russell, led to the construction of a remarkable diagram, the Hertzsprung-Russell (H-R) diagram; for many years it has served astrophysicists as an observational check for their theories of internal stellar structure and has become a

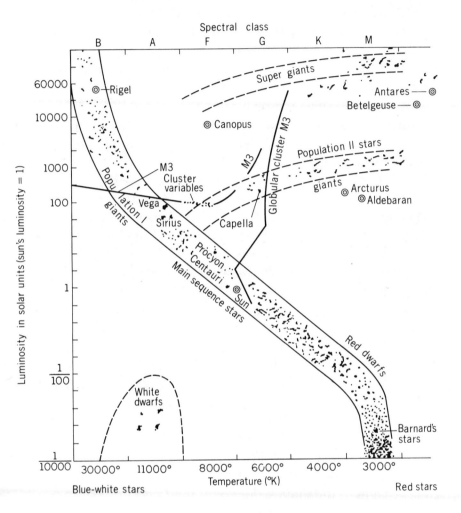

The Hertzsprung-Russell diagram. The thick solid lines running from the main sequence to the giant branches represent stars of the globular cluster M3. Every point on this diagram represents a star whose luminosity (in solar units) is given by the number directly to the left of it on the vertical axis; its surface temperature is given by the number directly below it on the horizontal axis at bottom; and its spectral class is indicated by the letter above it on the upper horizontal axis.

powerful tool for the study of the evolution of stars. Indeed, it was this diagram that gave the first definite clue about the way stars evolve as they grow older.

The H-R diagram is a simple two-dimensional graph on which the luminosities of the stars are plotted against their colors or surface temperatures; this is why this diagram is also called a color-luminosity diagram. The luminosities are plotted along the vertical axis of the graph and the surface temperatures along the horizontal axis. A single star is shown as a single point on this diagram; the horizontal position of the point shows the surface temperature of the star and the vertical position of the point shows its luminosity.

Suppose now that a few thousand stars are chosen at random over the entire sky and are placed on the diagram. One might suppose that the points do not fall into any specific pattern but instead are spread out randomly, but this is not so. Hertzsprung and Russell discovered that the stars lie along a few well-defined separate tracks whose significance for the evolution of stars was not clearly understood until quite recently. In the original work of Hertzsprung and Russell, only two tracks were shown in the diagram, but it has been shown that stars fall along four major tracks when they are plotted on the H-R diagram. About 90 percent of the stars lie along a fairly broad, elongated S-shaped band that extends from the upper left-hand region of the diagram to the lower right-hand region. This band of stars, which contains stars like Rigel, Sirius, Vega, the sun, and the orange and red dwarfs, is called the main sequence. Above and extending upward and to the right of the main sequence is an almost horizontal band consisting of yellow and red luminous stars that are, on the average, about 100 times more luminous than the sun; these are the giants. Still higher up are the superluminous red and yellow supergiants, which form still another upwardly tilted horizontal branch. The white dwarfs lie in the lower left-hand part of the diagram along a track that is almost parallel to the main sequence.

6 Birth, Evolution, and Death of Stars

The H-R Diagram of Star Clusters

All attempts to unravel the clues to the evolution of stars contained in the Hertzsprung-Russell diagram were unsuccessful until the theory of the internal structure of stars showed two important things: (1) the structure of a star at any moment in its history is determined by its mass and its chemical composition (essentially, the amount of hydrogen and helium it contains); (2) the chemical composition of a star changes as it ages, because it generates energy by fusing hydrogen into helium, thus altering its chemistry. This means that the position of any star in the H-R diagram is determined by three numbers: its mass, its chemical composition, and its age. Now the H-R diagram of a few thousand nearby stars chosen at random certainly includes a large range of masses, a variety of chemical compositions among stars at their birth, and a great variety of ages. The H-R diagram of a group of such stars presents no simple clues as to the evolutionary tracks along which points in the H-R diagram move as the stars that they represent age.

But astronomers discovered a way out of this difficulty when they began to study the physical properties of stars that live together and move through space together in clusters. Besides globular clusters, which consist of population-II stars and form a halo around the Galaxy, there are star clusters called open clusters, or galactic clusters, which lie in the plane

135

of the Galaxy and consist, in general, of a few hundred population-I stars intermixed with gas and dust that form a loosely knit gravitational structure. In some instances they contain as many as a few thousand stars or as few as 30 or 40. All the stars in such clusters move together around the core of the Galaxy, showing that they form a single generic family and were born together from the same chemical mixture. The stars in both open and globular clusters display wide ranges of luminosities and colors and this is the clue to the evolutionary track along which a star travels in the H-R diagram as it ages. Since all the stars in a cluster were formed from the same chemical mixture, they all had the same initial chemical composition and each star was chemically homogeneous throughout. Moreover, all the stars in a cluster are the same age because they were all formed at the same time from the original cloud of dust and gas. These two statements eliminate from consideration two of the numbers—the chemical composition and the age—that determine the position of a star in the H-R diagram and leave only the initial mass of the star to deal with. One may state this somewhat differently, and perhaps more clearly, as follows: The difference in the positions in the H-R diagram of two stars belonging to the same star cluster can only be caused by the difference in their initial masses, since their initial chemical compositions were, and their ages are, the same. In other words, the H-R diagram of the stars in a single cluster presents a clear picture of how the initial mass of a star affects its evolution in a given period of time.

To see how the study of the H-R diagrams of both open clusters and globular clusters has given us an empirical insight into the evolutionary tracks of stars in the H-R diagram, consider the various H-R diagrams of such clusters shown here. Among the diagrams are that of the very old open cluster M67 and that of the globular cluster M3, the famous and beautiful globular cluster in Canis Venatici. These diagrams show that every cluster, open or globular, has many

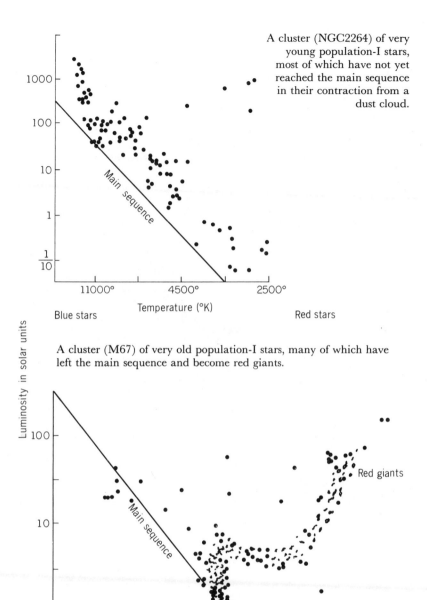

A cluster (NGC2264) of very young population-I stars, most of which have not yet reached the main sequence in their contraction from a dust cloud.

A cluster (M67) of very old population-I stars, many of which have left the main sequence and become red giants.

H-R diagrams of a very young cluster of stars *(top)* and a very old cluster *(bottom)*.

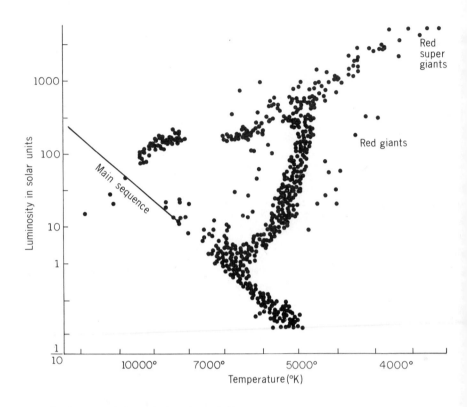

H-R diagram of the globular cluster M3.
   In the old globular cluster, M3 in Canes Venatici, the most massive stars have evolved to the red supergiant branch; stars of low mass are still on the main sequence.

stars that lie along the main sequence and some stars that lie along a branch that turns upward and to the right of the main sequence. The clusters M67 and M3 have well-developed branches that extend from the main sequence up into the red-giant and supergiant regions of the H-R diagram and then back down again. Since the stars in any one cluster are all the same age and started with the same initial chemical composition, these different branches must represent different stages in a star's evolution as determined by its mass. If one accepts this as a fact, one is faced with the question, On which branch of the H-R diagram does a star begin its life—on the main sequence or on the red-giant branch? The answer to this question is found in the theory of the internal structure of stars, which astrophysicists have fashioned into a remarkably powerful analytical tool for taking a star apart mathematically and seeing exactly how it is built and operates. This type of analysis has shown us that a star that is chemically homogeneous throughout must lie on the main sequence; such stars cannot be red-giant stars or be off the main sequence. The mathematical analysis shows that red-giant stars cannot be chemically homogeneous structures; they are chemically inhomogeneous, with dense helium cores and hydrogen-rich shells. Thus, when stars are formed from the gravitational condensations in a homogeneous mixture of gas and dust, they begin their active lives as stars on the main sequence because they are chemically homogeneous structures. For this reason the main sequence is also called the zero-age line of stars. After spending some time on the main sequence, the stars become chemically inhomogeneous as they transform hydrogen into helium in the core and evolve into red-giant stars and beyond. The more massive a star is, the faster this happens.

This picture of the way stars evolve and change as they grow older is one of the most beautiful examples of the way observational astronomy and theoretical astronomy or astrophysics, the field of astronomy that deals with the theory of stellar structure, have collaborated to reveal a sequence of events in the lives of stars that spans billions of years. The empirical analysis of the evolutionary data from the H-R diagrams of open clusters permits one to deduce the ages of clusters from the diagram in a very ingenious way. The older a cluster is, the larger is the number of stars in it that have evolved away from the main sequence and the more extensive is its non–main-sequence branch. This means that in old clusters the point along the main sequence where the stars begin to turn away from the main sequence is lower than for young clusters, as shown in the composite H-R diagram of various open clusters. One more fact is easily deduced from the H-R diagrams of star clusters. Since all the stars in the cluster began their lives on the main sequence, those that are now red giants must have started out as massive, luminous blue-white giants high up on the main sequence and moved from left to right in the H-R diagram as they evolved.

Evolution to the Main Sequence

An examination of the H-R diagrams of the globular cluster M3, which consists only of metal-poor, population-II stars, and of the open cluster M67, which contains only metal-rich, population-I stars, reveals that, in a general way, the evolution of both stellar populations is similar in the sense that they both evolve away from the main sequence toward the red-giant region. But the evolutionary details are different because the initial chemistries of the stellar populations were different. The population-II stars began their lives when the universe—and hence their composition—consisted of about 75 percent hydrogen and 25 percent helium, with traces of carbon and nitrogen, whereas the population-I stars began with 3 to 4 percent heavy elements and about 71 percent hydrogen. Although this difference in the chemical composition appears to be small, it has an important bearing on the evolution of these stars; the two stellar populations are therefore considered separately, but the way clouds of gas become main-sequence stars is the same for both populations.

Since all stars, regardless of the population to which they belong, were formed by gravitational contraction from clouds of gas or clouds of gas and dust, the events of their very early lives were about the same except for the duration of infancy, which depends on the star's mass. The more massive the material that was to become a star was when it began to contract, the more rapidly did the embryonic star pass

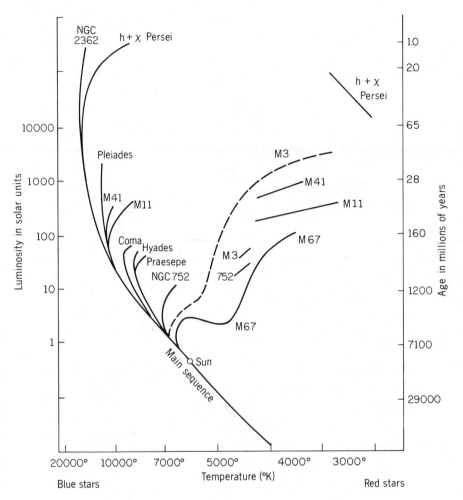

Composite H-R diagram of a number of different open galactic clusters and the globular cluster M3. The ages of the clusters are given along the vertical axis on the right. The farther down on the main sequence the turn-off point of a cluster is, the older it is. Thus M67 is the oldest cluster shown and NGC2362 is the youngest.

through its infancy and become a main-sequence star, and the higher up on the main sequence did it land when it began the stable part of its life. Thus, a contracting cloud whose mass was only 8 percent of the sun's mass took about 800 million years to reach the zero-age main sequence, whereas a contracting cloud whose mass was equal to the sun's mass reached the main sequence in about 10 million years. If the initial mass of the cloud was three or more times the sun's mass, the contraction time was only a few hundred thousand years.

Because of the similarity in the pre–main-sequence contracting phase of the lives of stars in general, discussion here will be limited to the description of how the sun, or a star like it, contracted from a cold cloud of gas and dust to its present, hot configuration. Contraction starts quite slowly but accelerates rapidly as the initial cloud collapses down to a radius that is about 100 times the sun's present radius. As it does so, and this takes no more than a few hundred years, it releases vast quantities of gravitational energy, half of which is radiated away into space in the form of ordinary electromagnetic radiation (light and heat) and half of which remains within the protostar as heat (the internal energy of the random motions of the atoms and molecules). This internal energy raises the interior temperature of the collapsing protostar to such an extent that the internal pressure slows the collapse down appreciably and the subsequent contraction proceeds in a slow, stable manner.

At this stage of its infancy the protostar is a very luminous cool sphere and looks very much like a red giant; it has a surface temperature of about 3,000°K and emits about 700 times as much energy every second as the sun does now. The chances are practically zero, however, that any of the red giants now visible are stars in their infancy. The reason is that protostars pass through this phase of their infancy so fast that hardly any are around for man to view. Moreover, if one of the typical red giants were a contracting protostar, astronomers would easily have detected definite changes in its appearance during the last century.

Once the protostar has contracted down to a sphere whose radius is about 60 times the sun's radius, after about

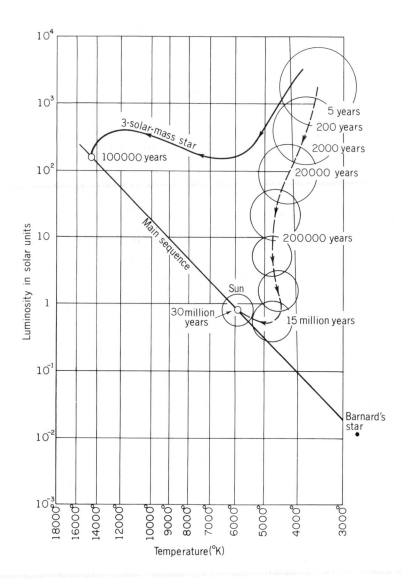

The evolution of stars onto the main sequence as shown by their tracks in the H-R diagram.

The descent of a star like the sun and that of a star of 3 solar masses onto the main sequence in the H-R diagram. The star of 1 solar mass (any star like the sun) contracts from an orange giant to a yellow main-sequence star in about 30 million years. The contraction in size to that of the present sun is shown.

The upper solid line shows the descent of a star of 3 solar masses which contracts to the main sequence in about 100,000 years.

1,000 years, its luminosity is about 500 times that of the present sun; its contraction then proceeds quite slowly. During this period the interior of the protostar is in a churning state (convection) in which the hot material from the deep interior is constantly dredged up in the form of vast currents and mixed with the cooler outer regions and cooler material sinks toward the center, so that the entire structure remains homogeneous. After about one million years of slow contraction, the radius of the contracting protostar is about twice that of the present sun and its luminosity about 1.5 times as large as the sun's. During this million-year period of contraction, the temperature in the deep interior becomes high enough for thermonuclear reactions to set in, and such light nuclei as deuterium, lithium, beryllium, and boron are transformed to helium; at the same time a good deal of the carbon is transformed into nitrogen, thus establishing the presently measured ratio of carbon to nitrogen in the sun. Contraction continues as the protostar becomes much less luminous and prepares to become a main-sequence star; after contracting for another 20 or 30 million years, it finally settles onto the main sequence and begins its life as a normal main-sequence star which may be called the initial, or zero-age, sun. The temperature at the center of the initial sun was about 10 million degrees, which is high enough to trigger the proton-proton chain of thermonuclear transformations. This consists of a series of steps during which four protons are fused into a helium nucleus and energy is released. This stage of contraction of the protostar is considered as the initial sun because gravitational contraction has ceased at this point and energy is released entirely via thermonuclear fusion; up until this point most of the energy released by the contracting protostar came from the gravitational contraction.

The initial sun was about 4.5 billion years younger than the present sun and 38 percent less luminous; its radius was 4 percent smaller than that of the present sun and its surface temperature was about 10 percent lower than the present sun's. In other words, the initial sun was a chemically homogeneous cooler, redder, smaller, and less luminous gaseous sphere than the present sun, which, with a central temperature of about 15 million degrees, is already chemically inhomogeneous in its central core because of the transformation of hydrogen to helium and is beginning to evolve away from the main sequence.

The description of how a population-I star like the sun contracts from a cloud to its zero-age main-sequence stage applies as well, with some variations in certain specific details, to population-II stars and to population-I stars more massive than the sun. In the case of massive stars, the descent to the main sequence is much more rapid than the sun's descent and the initial main-sequence star is higher up on the main sequence—bluer, hotter, and more luminous than the initial sun. Its central temperature ranges from 20 to 30 million degrees, which is hot enough to keep the carbon-nitrogen cycle going. This cycle releases energy by a series of nuclear transformations involving hydrogen, carbon, and nitrogen that fuses four protons into helium.

The Evolution of Population-II Stars

Since the population-II stars are the oldest stars in the universe and were formed from the primordial gas following the big bang we shall discuss their evolution first. Another reason for discussing the evolution of population-II stars before population-I stars is that the latter could never have been born if population-II stars had not existed and evolved first, for it was in the deep interiors of the very oldest population-II stars, near the ends of their lives, that the heavy elements required for population-I stars were built up from hydrogen and helium by thermonuclear fusion.

Since much of what is said here about the very early evolution of population-II stars of solar mass applies to population-I stars also, the sun will be used as an example and point of reference, even though the sun is already evolving away from the main sequence. As noted, when a star like the sun begins its normal life as a main-sequence star, its gravitational contraction stops and its central temperature is high enough to ignite the thermonuclear fire that slowly transforms hydrogen into helium. This occurs in the following series of steps:

(1) $$H^1 + H^1 \rightarrow H^2 + e^+ + \nu$$

(2) $$H^2 + H^1 \rightarrow He^3 + \text{radiation}$$

(3) $$He^3 + He^3 \rightarrow He^4 + 2H^1 + \text{radiation.}$$

In the first step, two protons (H^1) coalesce to form a nucleus of heavy hydrogen (H^2), also called a deuteron, with the emission of an antielectron or positron e^+ and a neutrino ν. In the second step the deuteron, H^2, captures a proton, H^1, and a helium-3 (He^3) nucleus is formed. He^3 is a light isotope of helium whose mass is 3 atomic units so that it is one unit less massive than the ordinary helium-4 (He^4) nucleus. In the third step, two He^3 nuclei coalesce to form an He^4 nucleus and to release two protons. Altogether these three steps—actually five steps, because each He^3 nucleus requires two steps for its formation and two He^3 nuclei are needed in step 3—result in the fusion of four protons to form a single He^4 nucleus. This process releases energy according to Einstein's famous relationship $E = mc^2$, because the mass of an He^4 nucleus is smaller than the mass of the four protons that coalesce to form it. In fact, the mass of four protons is 4.0325 atomic mass units, whereas the mass of an He^4 nucleus is 4.0039, so that 0.0287 units of mass are converted into energy every time four protons coalesce to form a single He^4 nucleus. Expressed somewhat differently, when 4.0325 grams of hydrogen are transformed into 4.0039 grams of ordinary helium inside stars like the sun, 0.0286 grams of mass are converted into energy. The amount of energy thus released is 600 billion calories— enough to melt 8 billion grams, or almost 20 million pounds, of ice. Not all the energy released in this proton-proton chain contributes to the luminosity of the star, because a small amount is carried off by the two neutrinos released, which pass right through the sun and out into space. The two positrons that are created do contribute to the star's luminosity because they are immediately annihilated when they meet two ordinary electrons.

This thermonuclear fusion of hydrogen into helium proceeded extremely slowly in the deep interiors of zero-age main-sequence stars like the sun or zero-age population-II stars. Indeed, even now any single proton, on the average, may wander around for billions of years inside the sun before it is caught in the thermonuclear dance that binds it with three other protons to form a helium nucleus. The reason for this is that the probability that the first step in the proton-proton chain will occur is extremely small because it involves the creation of a positron which can occur only when one of the protons in step 1 changes into a neutron. In spite of the slowness of the proton-proton chain, it is the prime thermonuclear process that accounts for the luminosities of the cool main-sequence stars. The sun is radiating at its present rate because in its deep interior about 564.5 million tons of hydrogen are fused into 560 million tons of helium every second via the proton-proton chain. Thus, to keep itself going, the sun is converting 4.5 million tons of mass into energy every second in spite of the slowness of the proton-proton chain; and the reason for this is that there are such vast quantities of hydrogen (protons) in each cubic centimeter in the deep interior of the sun that very large numbers of them are undergoing fusion at any given moment. In fact, there are about 50 trillion trillion protons in each cubic centimeter of the core of the sun, where the proton-proton chain is operating. Since this core contains about 30 million trillion

trillion cubic centimeters, some 1,000 trillion trillion trillion proton-proton fusion chains must be going on every second inside the sun, even if any single proton meanders around for billions of years before it is captured by another proton to start its transformation to helium. In spite of the fact that the sun and stars like it are converting 4.5 million tons of mass into energy every second via the proton-proton chain, these stars are so massive that the process would diminish their masses by only one-tenth of a percent in 15 billion years. Hydrogen is thus a very long lasting fuel for such stars.

In addition to the proton-proton chain defined by the three steps, there are alternate, but much less probable, chains that lead to He^4. These differ from the above chain in step 3 in that an He^3 nucleus combines with an He^4 nucleus to form a beryllium nucleus instead of combining with another He^3 nucleus to form He^4. The final result, however, is the same, since the beryllium nucleus quickly captures another proton to form a boron nucleus, which is unstable and quickly breaks up into two He^4 nuclei. Another important set of nuclear reactions that does not play a role in cool main-sequence stars like the sun but is of prime importance in the hot main-sequence stars is the famous carbon-nitrogen (CN) cycle, which also leads to the fusion of four protons into He^4. Since this cycle requires carbon for its operation and the initial population-II stars contained very little carbon, discussion of the CN cycle will be deferred. This cycle is important only when the core temperature of a star is 20 million degrees or higher and these high temperatures are found only in massive stars or stars that have evolved away from the main sequence.

Although the similarity between the sun and the early main-sequence life of a young population-II star is sufficiently great to allow comparison, as time goes on important differences develop. The precise details of the way a population-II star evolves away from the main sequence depends on the amount of He^4 that was present initially. If one accepts 25 percent as the concentration of He^4 in the initial main-sequence population-II stars, one finds that calculations of evolutionary models give evolutionary paths that agree fairly well with the paths deduced from the H-R diagrams of globular clusters. As initial population-II stars, whose masses were similar to the sun's, burned their hydrogen, they moved slowly up the main sequence, becoming hotter, bluer, and more luminous. These stars began to turn off the main sequence, moving upward and slightly to the right when their luminosities increased by about a factor of 10 and their surface temperatures increased by 15–20 percent. The length of time the initial population-II star spends on the main sequence depends on its mass. Thus, a population-II star whose initial mass was about 0.75 solar masses spent about 12 billion years on the main sequence and then evolved quite rapidly—in less than a billion years—to the red-giant region.

The population-II star began to turn off the main sequence because the thermonuclear fusion of hydrogen in its core transformed the star from a chemically homogeneous, to a chemically inhomogeneous, structure. This turn off was very slow and gradual at first but proceeded quite rapidly toward the giant branch of the H-R diagram when the star had exhausted all the hydrogen in its core, which then consisted of He^4 only. The star was then in its subgiant stage, during which it generated energy in a thin shell surrounding the helium core via the proton-proton chain. No nuclear energy was generated in the helium core itself during the subgiant stage, because the temperature there was far too low to involve He^4 in any kind of nuclear burning. But the hydrogen-exhausted helium core played an important role in the evolution of population-II stars through the subgiant stages of their lives because, since no nuclear energy was generated in the core, there was nothing to prevent the core from contracting gravitationally, with the result that the temperature of the core rose rapidly and large quantities of gravitational energy were released. Both of these phenomena caused the star's radius and its luminosity to increase quite drastically. The increase in the luminosity stemmed from the additional gravitational energy generated in the core and from the increase in the rate of thermonuclear fusion produced by the higher temperature in the hydrogen-burning shell surrounding the helium core. As the temperature increased to over 20 million degrees in the core, and hence in the shell, the CN cycle began to operate in the shell, thus increasing still further the rate of energy release and the luminosity of the star in its subgiant stage, even though the amount of carbon that was present initially was very small.

The rapid increase in the amount of energy generated in the hot hydrogen-burning shell and the contracting helium core caused the star to expand, and as it did so, its surface cooled off. Thus, the star became larger, redder, and more luminous as it moved almost vertically upward on the H-R diagram, reaching the giant branch in about one billion years. At this point, owing to the continuous contraction, the temperature in the core rose to 100 million degrees and a sudden and drastic change occurred in the core. The He^4 nuclei in the very hot core began to fuse with each other in the first step that leads to the triple helium reaction and to the formation of carbon-12 (C^{12}). Although a temperature of at least 10 million degrees is required to impart enough speed to protons to ignite the proton-proton chain, it is far too low to cause He^4 nuclei to fuse and form heavier nuclei. The reason for this is that although a temperature of 10 or 15 million degrees is large enough to overcome the mutual electrostatic repulsion between two He^4 nuclei, which is larger than that between two protons, and to bring them close enough to each other for nuclear fusion to occur, it does not occur, except momentarily, because the beryllium nucleus that is formed is not stable; it breaks down immediately into the original two He^4 nuclei again. At 100 million degrees, however, a third He^4 nucleus enters the scene to change the whole picture. At this high temperature so many pairs of He^4 nuclei are interacting to form beryllium-8 (Be^8) that there are always many Be^8 nuclei around, even though any one of them exists for no more than a fraction of a second. Owing to this, to the very high temperature, and to the great abundance of He^4 nuclei some of the short-lived Be^8 nuclei that are formed fuse with He^4 nuclei (alpha particles) to form the very stable C^{12} nucleus and release energy. This is the triple helium, or alpha particle, reaction, which may be written as follows:

$$He^4 + He^4 \rightarrow Be^8$$
$$Be^8 + He^4 \rightarrow C^{12} + radiation.$$

A complete analysis of this set of thermonuclear reactions, which is initiated when the evolving star reaches the red-giant branch of the H-R diagram, shows that it is ignited almost instantaneously. For this reason the onset of the triple helium reaction is referred to as the helium flash. Although vast quantities of energy are released during the helium flash, it occurs so deep down inside the star and lasts for so short a time that it has very little influence on the overall luminosity of the star. Once the helium flash is over, the star settles down to an orderly burning of He^4 in the core. In the early stages of helium-burning, most of the star's energy comes from the hydrogen-burning shell surrounding the helium core, but in time the helium-burning core supplies most of the energy. This brief description of the helium flash and the triple helium reaction gives a good picture of what happened inside a population-II star about one billion years after it had left the main sequence and became a red giant.

The population-II star considered here had an initial mass of about three-fourths the sun's mass and an initial He^4 content of 25 percent, but the overall picture is the same for population-II stars whose initial masses ranged from about 1.2 to about 0.65 solar masses and whose initial helium content ranged from 10 to 35 percent. The principal difference was in the time it took these stars to go from their initial main-sequence positions to their evolutionary paths. The more massive the stars were and the greater their initial helium content, the sooner they evolved to the giant branch. Thus, a star with 10 percent initial helium and 1 solar mass would have required 20 billion years to reach the red-giant branch, whereas a star with 35 percent helium content and 0.75 of a solar mass would have required 12 billion years. In any case, these data, derived from theoretical evolutionary models of population-II stars, lead to the conclusion that the stars in globular clusters are the oldest objects in the universe and that they had initial masses not much different from the sun's mass and helium contents of some 25 percent. Stars much more massive than the sun are not found in globular clusters, because such stars must have evolved so far during the cluster's life span that they are now all burnt out and contribute little to the cluster's luminosity.

The formation of carbon in the core of population-II stars was the beginning of the formation of the heavy elements. As the star transformed more and more of its helium to carbon, a core of carbon developed at the center of the helium core and two things began to happen: the carbon at the interface began to fuse with helium to form oxygen-16 (O^{16}), and the carbon core began to contract because no nuclear energy was generated within the carbon core itself to support the weight of the core. In time, however, the temperature within the carbon core, owing to the core's contraction, rose to hundreds of millions of degrees and the carbon nuclei began to fuse in pairs to form magnesium; at the same time, many more oxygen nuclei were being formed from carbon and helium, and these oxygen nuclei were in turn fusing with other He^4 nuclei to form neon.

Although the full details of how still-heavier nuclei were synthesized have not been worked out, a good overall picture of events during these last stages of the star's life can be painted in broad strokes. As successive cores were formed at the star's center at each new stage of the buildup of heavy elements, the star's central temperature continued to rise because of the continued core contractions until it reached more than a billion degrees. At these successively higher temperatures the alpha particles were captured by the successively heavier nuclei until all the helium in the original helium core was exhausted and nuclei of such heavy elements as silicon, sulfur, argon, calcium, and iron were built up. This heavy-element buildup produced by the capture of alpha particles did not go beyond iron, because alpha particles cannot be captured by iron nuclei. When a very energetic alpha particle collides with an iron nucleus, the iron nucleus is disrupted into less massive nuclei and other alpha particles; buildup beyond iron does not occur. At this stage of the star's development a drastic change took place because all its nuclear fuel was exhausted and there was no way for it to replenish the energy that was rapidly streaming out of its interior. As the star's internal temperature dropped, its internal pressure also dropped, and a violent gravitational collapse occurred because the star could no longer support its own weight. This collapse generated a tremendous amount of gravitational energy, which sent vast shock waves outward through the star, causing it to explode violently and to become thousands, or even millions, of times as luminous as it was.

The exact details of this stellar outburst are not fully understood, but there can be no doubt that such violent events occurred and are still occurring in the lives of stars as they pass through old age. This conclusion is supported not only by the theory of the structure and evolution of stars but also by direct observational evidence presented by the explosions of novas and supernovas, exploding stars that suddenly flare up and become, respectively, thousands of times or millions of times as luminous as they were. This happens in a very short time and is accompanied by a vast outpouring of material from the star's interior. Indeed, as in the case of the Crab Nebula in Taurus, which was observed as a supernova by the Chinese in the year 1054, most of the stellar material surrounding the iron-rich core of the star may be blown away, revealing the very hot, dense, small core. This core star at the center of the Crab Nebula is a hot neutron star that is spinning 33 times per second and emitting intense radio pulses.

Although nuclei heavier than iron cannot be built up by means of alpha-particle capture, very heavy nuclei, such as those of bismuth, gold, lead, and uranium, can be built up from iron through the capture of neutrons that are emitted in copious quantities during the gravitational collapse and subsequent explosion of a supernova. One can now account for the chemical compositions of the metal-rich population-I stars like the sun. They all originated from the material ejected from novas or supernovas or from material ejected in some other way from old population-II stars. The famous planetary nebulas are examples of highly evolved population-II stars emitting material.

The Evolution of Population-I Stars

Since the gaseous matter emitted by exploding population-II stars consisted, in addition to hydrogen and helium, of such heavier elements as carbon, oxygen, calcium, and iron, these coalesced into heavy molecules, which in turn condensed in large numbers to form particles and grains of dust in cold interstellar space, just the way water molecules in the cold upper regions of the atmosphere condense to form raindrops, hailstones, and crystals of snow and ice. Such grains of dust accelerated the formation of population-I stars in two ways: First, owing to their large masses—much larger than the masses of individual molecules—they moved about very slowly and exerted a much greater gravitational pull on each other than did individual molecules; hence, they had a greater tendency than did molecules to collect gravitationally and form the nucleus of a protostar. Second, such dust grains prevented the hot radiation from the remaining core of the nova from passing through easily, so that this radiation pushed the grains closer together, thus increasing their tendency to coalesce gravitationally. These effects in dust clouds led to a relatively rapid formation of population-I protostars, as can be seen in the photographs of such clouds. In these photographs there are actual examples in the famous Herbig-Haro objects of such clouds breaking up into nodules and condensations that contract almost as we watch them. Evidence of this process of population-I star formation is seen in such clouds in the sun's neighborhood of the Galaxy. In the constellations of Orion and Taurus are found quite young population-I stars, which could not have coalesced from the dust and begun their main-sequence lives more than a few million years ago. In the Galaxy the metal-poor population-II stars considerably outnumber the metal-rich population-I stars, but the latter are of much greater interest than the former because they present a complete evolutionary tableau, there being young, middle-aged, and old population-I stars, and because these are the stars with planetary systems that can support life. These stars are called metal-rich stars only because the metal abundances in their atmospheres are much larger than in the atmospheres of population-II stars and not because these metal abundances are large compared to hydrogen and helium; even in these metal-rich stars the heavy atoms constitute only 2 percent of the total mass.

Once gravitational condensation begins in a dust cloud, the subsequent collapse and descent of the population-I protostar to the main sequence occurs quite rapidly, with the same general sequence of events as in the case of the population-II stars. One must now, however, distinguish between main-sequence population-I stars more massive, and those less massive, than the sun, because the energy-generating mechanisms are quite different in the two cases. In main-sequence population-I stars as massive or less massive than the sun, the proton-proton chain operates just as in the case of main-sequence population-II stars. But in massive main-sequence population-I stars, stars with masses greater than 1.5 times the sun's mass, the dominant energy-generating mechanism is the CN cycle first discovered in the late 1930s by the German-American physicist Hans A. Bethe. This cycle, which starts with the capture of a proton by a C^{12} nucleus, leads, in a series of steps, to the fusion of four protons into a helium nucleus and the reappearance of the carbon nucleus. Carbon thus acts as a kind of nuclear catalyst. The following are the steps that comprise the CN cycle:

$$C^{12} + H^1 \rightarrow N^{13} + \text{radiation}$$
$$N^{13} \rightarrow C^{13} + e^+ + \nu$$
$$C^{13} + H^1 \rightarrow N^{14} + \text{radiation}$$
$$N^{14} + H^1 \rightarrow O^{15} + \text{radiation}$$
$$O^{15} \rightarrow N^{15} + e^+ + \nu$$
$$N^{15} + H^1 \rightarrow C^{12} + He^4.$$

In this cycle two positrons (e^+) and two neutrinos (ν) are created; the two positrons are immediately annihilated by two electrons with the release of energy, but the neutrinos escape, contributing nothing to the luminosity of the star. Since both the carbon and nitrogen nuclei have large positive electric charges, they repel protons quite strongly. Hence, the CN cycle can operate only if the temperature is so high that the kinetic energy of the protons is high enough to overcome this repulsion; this is so if the temperature is 20 million degrees or more. Such high temperatures occur at the centers of main-sequence population-I stars whose masses are at least about 50 percent larger than the sun's mass. The CN cycle is then the dominant energy-generating mechanism; the proton-proton chain contributes relatively little.

Another important difference between the massive and less massive population-I stars is that there is a large convective core—streams of hot gases flowing away from the center and cool streams flowing in—in the former and no convective core in the latter. Since the highest temperature in a star is found at its center, the generation of energy by the CN cycle, which depends very strongly on the temperature, occurs almost entirely at the center. This means that the cores of upper main-sequence population-I stars are in intense turmoil because a tremendous amount of energy is released in a small region very close to the center and can only escape by being carried out by streams of hot gases. The convection keeps mixing the helium that is formed near the center of the core with the unused hydrogen in the rest of the core, so that the convective core remains chemically homogeneous. Thus hydrogen is depleted and He^4 is built up uniformly and quite rapidly throughout the entire core. In lower main-sequence population-I stars there is no convective core because the proton-proton chain that operates there does not depend very strongly on the temperature so that the energy generation is not confined to a small central region but occurs at an ever-decreasing rate with increasing distance from the center. Thus no convective core is built up, because the energy is radiated out to the surface quite easily. In a lower main-sequence population-I star the hydrogen content is greatest near the surface of the star and smallest near the center.

This difference in the chemical composition and behavior of the cores of upper main-sequence and lower main-sequence population-I stars accounts for the difference in the way they evolve away from the main sequence. The H-R diagram on page 158 shows in some detail the evolutionary track away from the main sequence of a population-I star whose initial mass was 5 solar masses; in the H-R diagram on page 159 the evolutionary tracks of a number of stars with a variety of different initial masses are shown. In these H-R diagrams $M_\odot$ stands for the mass of the sun and the designations $3M_\odot$, $1.25M_\odot$, and so on, mean 3 solar masses, 1.25 solar masses, and so on. These evolutionary tracks are based on the detailed mathematical analysis of Icko Iben, Jr., whose work is preeminent in this area.

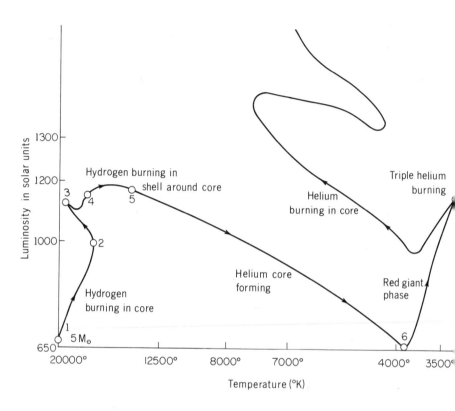

The detailed evolutionary track in the H-R diagram of a metal-rich star of 5 solar masses starting from the main sequence (1). This star uses up most of the hydrogen in its core in about 65 million years (going from 1 to 3). A helium core is formed in the next 2.5 million years (3 to 6) and in another half-million years it reaches the red-giant stage and the helium in the core is ignited (according to Icko Iben).

The difference between the ways a star whose mass is greater than $1.5M_\odot$ and one whose mass is less than that evolve away from the main sequence is caused by the existence of a convective core in the former and the absence of such a core in the latter. In the early main-sequence stages of the upper main-sequence population-I stars hydrogen burning proceeds quite steadily near the center for anywhere from some 10 million years for a star of $15M_\odot$ to about 2 billion

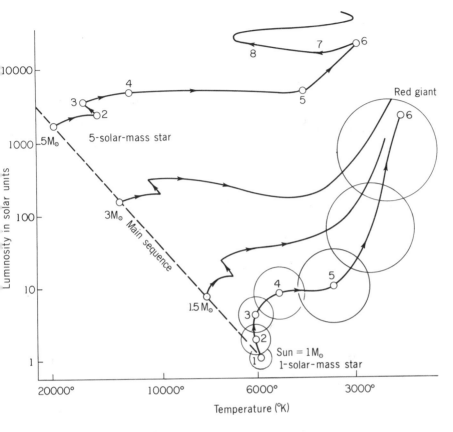

The evolutionary paths in the H-R diagram of four metal-rich stars with masses equal to 1, 1.5, 3, and 5 solar masses. The stars like the sun (1 solar mass) go from 1 to 2 in 7 billion years and from 2 to 5 in 2 billion years; they reach the red giant stage (6) in another 2 billion years. A star of 5 solar masses goes from 1 to 2 in 65 million years and from 2 to 5 in about 4 million years; the total evolutionary time of such a star from 1 to 6 (red giant) is about 70 million years (according to Icko Iben).

years for a star of $1.25M_\odot$. During this hydrogen-burning period, these stars grow cooler and more luminous, so that they move upward and to the right in the H-R diagram, as indicated by the numbers 1, 2 on their evolutionary tracks. The lower main-sequence population-I stars, on the other hand, spend billions of years burning hydrogen very slowly—

stars like the sun remain close to the main sequence for about 7 billion years—and become hotter and more luminous as they move upward and to the left.

When all the hydrogen is exhausted, or very nearly so, in the cores of the upper main-sequence stars, these cores contract and heat up quite rapidly; this produces an overall contraction of these stars, which causes their luminosities and surface temperatures to increase quite rapidly for a period that lasts anywhere from a few hundred thousand years for the more massive upper main-sequence stars to some tens of millions of years for the less massive upper main-sequence stars. Overall contraction goes on in any one of these stars until the temperature at the surface of the hydrogen-depleted core is high enough to ignite the CN cycle in a hydrogen-rich shell surrounding the core; the star then reverses its track in the H-R diagram becoming cooler, redder, and somewhat less luminous. This goes on until the mass of the helium core, now completely depleted of hydrogen, equals about 10 percent of the total mass of the star. At this point the mass of the core is so large that its rapid gravitational contraction is unavoidable. As the core contracts, its temperature rises quite rapidly and the star expands into the red-giant region, becoming redder, much larger, and much more luminous. A star whose initial mass is $5M_\odot$ expands until its radius becomes 74 times the sun's radius. At this point the core temperature is about 100 million degrees and the triple helium process $(3He^4 \rightarrow C^{12})$ is ignited in the core. The helium burning that thus goes on in the core contributes to the luminosity, ultimately becoming the main energy source of the star in its red-giant stage.

The evolution away from the main sequence and into the giant branch of lower main-sequence population-I stars with masses equal to, or less than, the sun's mass proceeds much more slowly and uniformly than that of the upper main-sequence stars. In fact, the evolutionary tracks of these stars are quite similar to those of the very old population-II stars. Since the lower main-sequence stars have no convective cores, they suffer no sudden overall contraction as hydrogen is being depleted. They first move leftward almost parallel to the main sequence for 7 or 8 billion years, becoming bluer, hotter, and more luminous, and then turn gradually to the right as the hydrogen in the core is slowly depleted. When hydrogen depletion is complete, the core temperature rises as the core contracts, and the star rapidly expands into the red-giant branch. At this point helium is ignited in a flash as the core temperature reaches 100 million degrees, and the energy thus released causes the core and the entire star to expand. Helium burning then proceeds at a steady rate until enough C^{12} has been synthesized in the core for the $C^{12} + He^4 \rightarrow O^{16}$ reaction to begin and to set the stage for the subsequent buildup of the still heavier elements as the core temperature rises to hundreds of millions, and finally billions, of degrees. The physical processes that go on during this phase of the population-I star's life are quite similar to those that go on inside population-II stars that have passed through their triple-helium-burning stage and are building up heavier chemical elements by means of alpha-particle capture. Just as in the case of population-II stars, the population-I stars will end their lives by becoming novas after all the helium in their hot cores has been transformed into the various heavy elements up to iron. There is, however, one important difference between the buildup of heavy elements in population-II and population-I stars. Since population-II stars start out with almost all hydrogen and helium, elements whose atomic weights are not multiples of 4, such as fluorine-19 (F^{19}) and sodium-23 (Na^{23}), have little chance of being formed. The nuclei whose atomic weights are multiples of 4 are favored because 4 is the atomic weight of He^4, which plays the dominant role in heavy-element buildup. Population-I stars are formed from a chemical mixture that already contains heavy nuclei. Since these can capture protons in addition to He^4 nuclei, the restriction to nuclei whose atomic weights are multiples of 4 is removed. Thus, the oxygen nucleus can capture a proton to become an isotope of fluorine, the neon nucleus can capture a proton to become a sodium nucleus, and so on.

Note again the important—indeed, overwhelming—role that the initial mass of the star plays in its evolution away from the main sequence. In the H-R diagram of the evolutionary track of a $5M_\odot$ star the intervals of time between successive stages in the star's structure and behavior are indicated, and similar results can be derived from the theory of stellar structure and evolution for stars of larger and smaller mass. Without going into these details one must note the following: a $15M_\odot$ star evolves from the main sequence to the tip of the giant branch in about 12 million years; a $9M_\odot$ star evolves to the red-giant tip in about 25 million years; a $5M_\odot$ star takes some 90 million years to evolve; a $3M_\odot$ star takes about 300 million years; and a star as massive as the sun takes over 10 billion years to reach the red-giant tip. Roughly speaking, the time in years that a population-I star spends on the main sequence equals 10 billion multiplied by the mass of the star (in solar mass units) divided by its luminosity (in solar luminosity units).

The Evolving Sun

The general description of the evolution away from the main sequence of aging population-I stars as they transform their hydrogen to helium can be applied to the sun if the data for a star of one solar mass are used. The evolutionary track of such a star in the H-R diagram on page 159 is shown at the bottom of the main sequence in the composite figure of the evolutionary tracks of stars of different masses. From this it is seen that the sun is a very slowly evolving star that is just beginning to leave the main sequence. Its surface temperature is still increasing, so that it is becoming bluer, larger, and more luminous as it moves almost parallel to the main sequence. It will continue changing in this way, departing more and more from the main sequence in an upward direction for about 5 billion years. It will then turn sharply to the right and continue to move to the right for about 2 billion years, becoming much larger, considerably cooler, redder, and still more luminous. After this relatively quiescent stage it will move drastically upward as it expands into its red-giant phase and will become about a thousand times more luminous than it is now.

Since the sun is now almost 5 billion years old, it has already converted half the hydrogen in its core to helium, so that it is no longer chemically homogeneous. At this stage of the sun's development the density of matter at its center, the gaseous mixture of H^1 and He^4, is about 160 grams per cubic centimeter (compare this with the overall solar density of 1.44 grams per cubic centimeter), and the temperature at its center is 15 million degrees. At present the sun is converting mass into energy at the rate of 4.5 million tons per second, but this rate will increase as the central temperature and density of the sun rise in the next few billion years. This will in turn cause the sun to become still hotter, somewhat bluer, and more luminous. It is obvious that this will greatly affect the earth and life on it, just as the lower surface temperature and lower luminosity of the sun about a billion years ago greatly affected the emergence of life on the earth then. Since the sun's luminosity was about 10 percent smaller a billion years ago, the earth's overall surface temperature could not have been higher than 0°C, so that life could not have existed. Using the same kind of argument in reverse, one can see that the hotter, more luminous sun that will dominate the solar system in a billion years will drastically alter, or entirely destroy, life on the earth. The sun's luminosity will have doubled by that time and the overall temperature of the earth will have increased by at least 50 percent, so that some rivers and lakes will begin to boil.

7 The Origin of the Solar System

The electromagnetic force plays only a minor role in the life histories of stars but is quite important in those of the planets and the other bodies in our solar system. The nuclear force was negligible in the processes that led to the formation of the planets; but the gravitational force was still important, and the electromagnetic force played a dominant role in giving the planets their present properties. Gravity was very important in bringing together the bits of matter that formed each planet, but the final step was governed and dictated by the electromagnetic force because all chemical reactions are governed by this force and chemical processes in the past finally determined how each planet would look and behave today.

The Sun's Planets

When Giordano Bruno (1548–1600), Italian philosopher and disciple of Copernicus, was burned at the stake as a heretic, it was not only for expanding and supporting the Copernican heliocentric theory but also for teaching his own much more revolutionary doctrine that each star is a sun attended by a retinue of planets. This very bold and far-reaching concept remained a mere speculation until astrophysicists demonstrated in recent years that stars are born from clouds of dust and gas by gravitational contraction. An analysis of the steps that lead from an amorphous dust cloud to a star shows that planets must be as natural a consequence of these steps as the star itself is. One may infer from this that Bruno was, indeed, correct.

Even without the evidence presented by the theory of stellar formation described in the previous chapter, one must conclude from the regularities and symmetries found in the physical and dynamical properties of the planets that the sun and planets were formed together. The solar system is clearly no accidental arrangement of bodies but an ordered structure whose remotest planet is no farther away from the sun than the six-thousandth part of the distance to the nearest star; all bodies in the solar system accompany the sun in its rapid motion around the center of the Galaxy. As viewed from a distance of a trillion miles or so, the solar system looks like a thin rotating disk of very tenuous matter surrounding a hot, luminous sphere, the sun. From a still greater distance one would see that this entire structure is surrounded by a thin shell or halo of loosely structured bodies, the comets. The thin disk surrounding the central hot body consists of various kinds of gases, dust, particles, and debris, all swirling in the same direction around the central gaseous sphere, with the outer parts of the disk revolving more slowly than the inner parts. In addition to the dust and gaseous material, the disk contains nine fairly large spheres of solid matter—the planets—spaced at varying distances from the disk's center and all circling around the center in the same direction but at different speeds, with the innermost of these solid spheres revolving more rapidly than the outermost. All but the outermost one and the innermost two of these nine spheres have small spherical satellites, or moons, revolving around them.

The disk just described is perhaps the most striking feature of the solar system, for it indicates more clearly than does anything else the symmetry of the sun's planetary system and the strong relationship between the origin of the sun and the birth of the planets themselves. If the planets were no more than interlopers from distant space that had acciden-

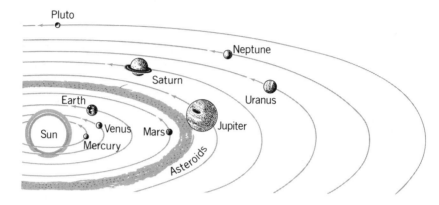

The disk of the solar system. The planets and asteroids all revolve around the sun in the same direction.

tally wandered into the gravitational field of the sun and been captured by it, the arrangement of the planets would not be so orderly. To emphasize this point, the following planetary features are noted: All the planets revolve around the sun in the same direction, in elliptical orbits whose planes almost coincide. Except for Mercury, the planet closest to the sun, and Pluto, the most distant planet, the planets move around the sun in orbits that lie in planes that are tilted by no more than 3.5° with respect to the plane of the earth's orbit, called the plane of the ecliptic. Although the tilt of the plane of Mercury's orbit is about 7° and that of Pluto's orbit is 17°, the overall picture is that of a system of orbits that lie almost in the sun's equatorial plane.

There is also considerable regularity in the distances of the planets from the sun, which is most clearly revealed in a numerical rule for the distances that was first discovered by the German astronomer Johann Daniel Titius in 1776 and later confirmed by another German astronomer, Johann Elert Bode (1747–1826), who was unaware of Titius' discovery. This Bode-Titius relationship, or numerical rule, for the distances states that, starting with Mercury, the mean distances of the planets from the sun, expressed in terms of the earth's mean distance (the astronomical unit, which is taken as 1), are as follows:

PLANET	DISTANCE	PLANET	DISTANCE
Mercury	0.4	Jupiter	0.4 + 16(0.3)
Venus	0.4 + 0.3	Saturn	0.4 + 32(0.3)
Earth	0.4 + 2(0.3)	Uranus	0.4 + 64(0.3)
Mars	0.4 + 4(0.3)	Neptune	0.4 + 128(0.3)
asteroids	0.4 + 8(0.3)		

It is important to note in connection with this rule that it deals with the mean distances of the planets from the sun; since the planets move in elliptical orbits around the sun, their distances from the sun vary from moment to moment. A planet's closest approach to the sun is called its perihelion and the point of greatest distance from the sun is called its aphelion; its mean distance from the sun is half the sum of its perihelion and aphelion distances. Since the earth's perihelion distance, measured in December, is 91.5 million miles and its aphelion distance, measured in June, is 94.5 million miles, the earth's mean distance—the astronomical unit—is half of 186 million miles, or 93 million miles. When Bode first stated the rule for the planetary distances, the asteroids, irregularly shaped bodies of various sizes and masses, thousands of which form a belt between Mars and Jupiter, had not yet been discovered, so that the rule appeared to be flawed by a gap between Mars and Jupiter; but this gap was filled on January 1, 1801, when the Italian astronomer Guiseppe Piazzi (1746–1826) discovered Ceres, the largest and most massive of the asteroids, which are also called the minor planets. The mean distance of Ceres, whose diameter is about 400 miles and whose mass is one ten-thousandth of the earth's mass, is almost exactly equal to the value given by Bode's law. The asteroids seem to be the components of a planet whose mean distance would have been exactly that given by Bode's law if it had ever been formed. Although for many years Bode's law

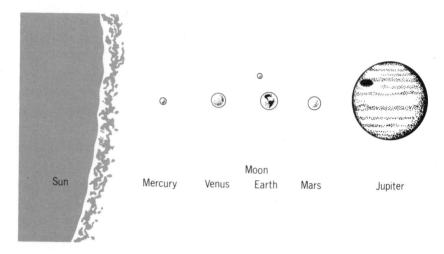

Sun   Mercury   Venus   Moon
Earth   Mars   Jupiter

was thought to be a numerical coincidence, today it is believed to be a real law that is deducible from the dynamics of the origin of the solar system. In fact, the most recent and complete theories of the birth of the planets lead to expressions for the mean planetary distances that are similar to Bode's law. In any case, the regularity of the mean distances of the planets from the sun very strongly supports the argument that the sun and planets were formed together.

The physical and chemical properties of the planets also show a remarkable regularity if they are first divided into two groups: the four inner, or terrestrial, planets, Mercury, Venus, Earth, and Mars, whose orbits lie within the asteroid ring; and the four Jovian planets, Jupiter, Saturn, Uranus, and Neptune, whose orbits lie beyond the asteroid ring. The four inner planets all have roughly the same masses, sizes, densities, and chemical compositions. Of these four planets' diameters the earth's is largest (8,000 miles) and Mercury's is the smallest (3,000 miles); the diameter of Mars is 4,200 miles, and that of Venus is 7,600 miles. The mass of the earth—about 6 billion trillion tons—is about 19 times the mass of Mercury, about 1.25 times the mass of Venus, and about 9

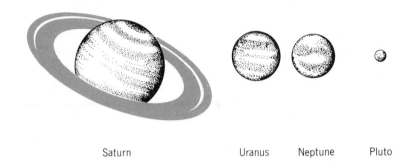

Saturn   Uranus   Neptune   Pluto

Relative sizes of the sun and the planets.

times the mass of Mars. The earth is the densest of these four planets, with a density of 5.52 grams per cubic centimeter, and Mars the least dense, with a density of 4 grams per cubic centimeter. Mercury's density is 5.5 grams per cubic centimeter, and that of Venus 5.1 grams per cubic centimeter. These mean densities, which are all larger than that of the ordinary silicate rocks (about 3 grams per cubic centimeter), indicate, as has been definitely established for the earth, that the cores of the inner planets consist of matter, probably iron or nickel, that is considerably denser than rock. The relative sizes of the planets as compared with the size of the sun are shown in the diagram above.

The inner planets as a group are marked by the small number of satellites revolving around them. Mercury and Venus have no satellites; the earth has one, the moon; and Mars has two very tiny ones, 5 miles and 10 miles in diameter, respectively. The moon revolves around the earth in an orbit that is tilted only slightly to the plane of the earth's orbit and, hence, to the earth's equatorial plane. The same thing is true of the two small satellites revolving around Mars: the planes of their orbits are tilted only about a degree with respect to the plane of Mars' orbit. Here again one sees a remarkable regularity in the dynamics of the inner planets, which points to a common origin of these bodies and the sun.

Not only are the planets revolving in elliptical orbits around the sun, but they are also rotating about axes that, except for the planet Uranus, are very nearly parallel to each other. The periods of rotation of the planets show a marked difference as one goes from the four inner planets to the Jovian planets. The inner planets are spinning relatively slowly; that is, their days are long. Mercury rotates on its axis once every 60 earth days; Venus appears to rotate in a reverse direction (retrograde rotation) once every 4 days, although this figure is still uncertain; and Mars spins on its axis once every 24 hours and 37.3 minutes.

Among these four planets, Mercury is the only one that has no atmosphere; it has too small a mass to hang onto the molecules of an atmosphere gravitationally. Venus has a hot, dry atmosphere that contains great quantities of carbon dioxide and nitrogen and is surrounded by an opaque cloud whose exact chemical nature is unknown. The pressure at the bottom of Venus' atmosphere may be about 100 times the earth's atmospheric pressure because of Venus' high surface temperature—about 800°F. This high surface temperature stems from the greenhouse effect caused by the great concentration of carbon dioxide in Venus' atmosphere; visible solar radiation passes quite readily through the atmosphere and reaches and heats the surface, but the heat rays emitted by the surface cannot escape because they are absorbed by the carbon dioxide molecules of the atmosphere. Mars has a very tenuous, hazy atmosphere with only small quantities of oxygen and water vapor in it; it appears to consist mostly of carbon dioxide. The total Martian atmosphere is about one-thirtieth of the earth's. The mean Martian surface temperature is about −65°F.

Jupiter, the largest and most massive planet in the solar system, has a diameter about 11 times the earth's diameter and a mass about 318 times the earth's mass; a person standing on Jupiter's surface would weigh about 2.7 times as much as he does on the earth. The mean density of Jupiter is about 1.33 grams per cubic centimeter, or 33 percent denser than water. This low mean density indicates that Jupiter is constructed mostly of very light atoms; a theoretical analysis of all the data shows that Jupiter's interior consists mostly of liquid and solid hydrogen, and its outer shell is a 15,000-mile-thick layer of ice. If Jupiter were much more massive than it is, it would not be a planet but a star because its central temperature would then be high enough to sustain thermonuclear fusion. Jupiter's atmosphere is composed mostly of hydrogen, helium, ammonia, and methane. Jupiter is surrounded by an intense magnetic field in which great quantities of very energetic electrons and protons are trapped.

Jupiter's mean distance from the sun is about 500 million miles, and it revolves around the sun once every 11.86 years. It rotates on its axis once every 9 hours and 50 minutes, so that its surface is quite flattened at the poles. Jupiter is attended by a retinue of twelve natural satellites that revolve around it in orbits that are either coplanar with Jupiter's equator (the first four or inner satellites) or tilted about 30° to that equator. Two of the four inner satellites are larger and more massive than earth's moon and one is as large and massive. We have considered Jupiter in some detail here because the three other Jovian planets are similar to it. Saturn, which, when viewed through a telescope, is one of the most beautiful objects in the sky because of its system of rings, is chemically similar to Jupiter and has the same internal structure and the same kind of atmosphere. Its mass is 95 times the earth's mass, and its diameter is about 9 times the earth's diameter so that its mean density is only 0.75 gram per cubic centimeter. Since this is less then the density of water, Saturn would float if it were in an ocean large enough to accommodate it. It rotates on its axis once every 10 hours and 20 minutes and revolves around the sun once every 29.5 years at a mean distance that is 9.5 times the earth's mean distance. Because of its rapid rotation and low mean density, Saturn is the most flattened at the poles of all the planets. Nine satellites revolve around Saturn.

Saturn, the most distant planet observable with the naked eye, was known to the ancients and probably figured prominently in their astronomy; because it takes 29 years to revolve around the sun, Saturn reappears in the same part of the sky for many years, appearing to swing from east to west once a year only because of the earth's revolution around the sun. The planet Uranus, the first of the three planets discovered in modern times, was initially observed quite by accident in the eighteenth century by Herschel while he was sweeping the sky with his telescope. Its diameter is about 4 times that of the earth, its mass is about 15 times the earth's mass, and its mean density is 1.6 grams per cubic centimeter. Although somewhat smaller and less massive than Saturn, Uranus is quite similar to it in chemical composition, internal structure, and atmosphere. There is probably very little ammonia in the atmospheres of Uranus and Neptune because the temperatures on these planets are so low that the ammonia is frozen. Uranus is accompanied by 5 satellites in its 84-year journey around the sun, moving in an orbit whose mean radius is 19 times the earth's mean distance. It rotates on its axis just about as fast as Saturn does, once every 10 hours and 49 minutes, and is about as flattened at its poles as Jupiter is.

The existence of the planet Neptune was deduced theoretically in the 1840s by the brilliant French astronomer Urbain Jean Joseph Leverrier (1811–1877) and, independently, by the British astronomer John Couch Adams (1819–1892). Both of these investigators decided that the very small observed deviations of Uranus from the orbit theoretically predicted for it from Newton's law of gravity were caused by the presence of a more distant planet. Almost simultaneously they both began to calculate, again using Newton's law of gravity, where such a planet would have to be to cause Uranus to depart from its predicted orbit, and both came up with the same answer at just about the same time. Their calculations predicted the presence of a planet like Uranus at a distance of about 30 astronomical units from the sun. Leverrier sent his predictions to the Potsdam observatory, and such a planet, Neptune, was picked up telescopically by the German astronomer Johann Gottfried Galle (1812–1910), the same night that he was informed about it, and very close to the predicted position. This discovery brought scientific theory to great eminence because it showed the enormous power of mathematical analysis when combined with a correct physical theory.

Neptune's size is slightly smaller than that of Uranus and its mass is somewhat larger, with a density of 2.25 grams per cubic centimeter. It spins on its axis once every 15 hours, so that it is not quite as flattened as Uranus, and revolves around the sun once every 165 years. It has two satellites, one of which is larger and about twice as massive as earth's moon. Chemically and structurally it is similar to Uranus, but considerably colder.

The most distant planet in the solar system—as far as is known—is Pluto, at a mean distance of 39.5 astronomical units, or almost 4 billion miles, from the sun; it revolves around the sun once every 248.5 years. This planet, whose existence was predicted by the American astronomer Percival Lowell (1855–1916) on the basis of the observed deviations of Neptune from its calculated orbit, was discovered in 1930 by another American astronomer, Clyde W. Tombaugh. Pluto's size, mass, and density are closer to those of Mars than to those of any of the other planets, so that in its physical characteristics, it is like the terrestrial planets and (except for the absence of oxygen) probably has an atmosphere like the earth's. Its period of rotation is 6 days and 9.5 hours. Since part of Pluto's orbit lies within the orbit of Neptune, some astronomers conjecture that Pluto may have been a satellite of

Neptune many years ago. If there are planets beyond Pluto, which may be the case, finding them will be extremely difficult because they are either very small or very far away.

The Remarkable Comets

Any theory of the origin of the solar system must account not only for the sun and its attendant planets but also for the comets, which are perhaps the most curious and amazing objects in the solar system because of the way they come and go and change their appearance. They differ in every respect from the planets, the satellites, and the asteroids. Instead of being compact, solid spherical bodies moving around the sun in fairly round orbits, they are loosely structured bodies that look like small disks when they first become visible and then develop long tails. They generally move in very elongated orbits around the sun when they come into the inner part of the solar system. These strange and mysterious interlopers have stirred more fear and been the object of more superstitious beliefs than any other kind of celestial body. For thousands of years they were considered as hairy (hence the name *comet* from the Greek *coma*, "hair"), evil objects, portending great disasters and misfortunes.

Careful and detailed analyses of the orbits of comets have revealed that these objects belong to the solar system; they are not strangers that wander by chance into the neighborhood of the sun from interstellar space. In no case has a cometary orbit been found that has the characteristics that are associated with the orbits of objects that do not belong to the solar system. The latter are hyperbolic orbits, whereas comets have elliptical orbits. In some cases comets have been observed to move in apparently hyperbolic orbits, but it has been definitely established that in such cases the orbits were initially elliptical and had been altered to hyperbolic orbits by the gravitational action of Jupiter or Saturn. Since the known comets usually move in very elongated orbits around the sun, they spend most of their time at such great distances from the sun that only rarely do men see the same comet return to the neighborhood of earth in their own lifetimes. From this it is deduced that there are billions of comets surrounding the solar system at a distance that is almost halfway to the nearest star. This conclusion is based on Oort's analysis of cometary orbits, which shows that the total number of comets that are observed circling the sun each year can be accounted for only if one assumes that they come from a vast cloud of about 300 billion comets that form a shell or halo around the sun at a distance of about 100,000 astronomical units. These comets move around the sun in extremely large, very nearly circular, orbits; most comets traverse their orbits once every 10 million years or so, but every year a few of them lose energy because of some disturbance (perhaps collisions with each other or the gravitational attractions of the closest stars) and fall in toward the sun along very flattened orbits. Certain comets, the periodic comets, are seen over and over again, because their orbits were foreshortened years ago by Jupiter's or Saturn's gravitational action. When one of the distant comets passes close to Jupiter or Saturn as it approaches the sun, its orbit is altered considerably; it may, indeed, be slowed down to such an extent that it does not have enough energy to return to the halo of comets from which it came, in which case it continues to move in an orbit around the sun that lies within Jupiter's or Saturn's orbit.

When a comet is very far away from the sun, it is probably a large compact ball of ice, snow, and dirt 10–20 miles in diameter. As this "dirty snowball," the nucleus of the comet, approaches the sun, it begins to melt and vaporize because of the sun's intense radiation, and a large coma of gases and dust is formed. This coma will, in general, become many times larger than a planet as the comet gets closer to the sun, becoming easily visible to an observer on the earth. Since the mass of a comet is exceedingly small, as indicated by the immeasurably small effect it has on the motion of a planet, it

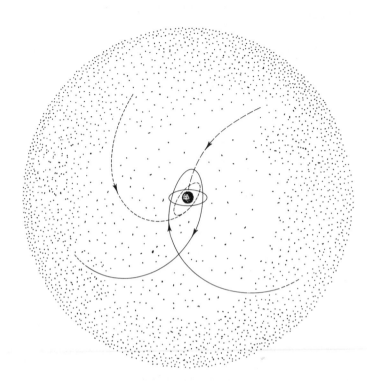

The vast halo of about 100 billion comets that circles the sun at a distance of about 2 light-years. Occasionally one of these comets falls into the inner region of the solar system, where it may be confined to a small orbit by the perturbing action of Jupiter.

cannot hold onto the gases and dust particles gravitationally that the sun's heat forces the nucleus of the comet to eject. Owing to this weak gravitational attraction between the core of the comet and the dust and gases in its coma, the solar radiation and the solar wind—a stream of high-energy ions ejected from the sun's surface—drive the dust and gases away from the comet to form a long tail, which is always directed away from the sun, whether the comet is approaching, or receding from, the sun. The length of the tail and its brightness depend on how large the head of the comet is, how much volatile material it contains, the intensity of the solar wind, and the comet's closest approach to the sun. When Halley's comet swept around the sun in 1910, it developed a tail that was about 100 million miles long, so that the earth passed right through its tail at one point in the earth's orbit. The tail, however, was so tenuous that it had no observable effect on the earth itself or on any living creature on it. The actual amount of material in the tail of a comet is so small that one can easily see the stars at night time shining through the tail.

Since some of the material in the head of a comet is driven away to form the tail each time the comet comes close to the sun in its perihelion passage, the mass of the comet is slowly diminished and the comet itself fades away after many trips around the sun. Indeed, some of the periodic comets whose orbits lie close to Jupiter's orbit have actually disappeared in this way; the material that was initially in the head of any such comet continues to move in the comet's orbit and contributes to the stream of meteors that continually strike the planets in the solar system. In certain instances, the tidal action of the sun may tear a comet into two, three, or four parts as the comet swings around the sun. These separate parts of the comet then continue to move together in very nearly the same orbit.

Meteors and Meteorites

An observer watching the sky on any clear moonless night can easily observe streaks of light flashing across the sky every now and then. The ancients actually interpreted these flashes as "falling stars," and this name is still used by many people in referring to such phenomena. This colorful description of these sudden flashes gives an entirely wrong impression of what they are; they are actually caused by meteors, particles usually no larger or more massive than grains of sand that enter the earth's atmosphere at a speed of about 20 miles per second and are immediately vaporized by the intense heat generated by the air's friction. The temperature of the meteor and of the air surrounding it increases to a few thousand degrees as the kinetic energy of the meteor is quickly transformed into heat; one then sees a sudden streak of light across the sky whose length, duration, and brightness depend on the size, mass, and speed of the meteor. If it were not for the earth's atmosphere, the meteors striking the earth would be very destructive to life, since their impact would have the same effect as a 45-caliber bullet.

The number of meteors that enter the earth's atmosphere in a single 24-hour period and are bright enough to be seen is extremely large. In fact, counts of meteors made by individual observers indicate that about 200 million visible meteors strike the earth's atmosphere in a single day, which means that about 100 tons of meteoric or cosmic dust settles on the earth's surface every day. Since the dust that is swept out of the head of a comet continues to move as a stream of matter in the orbit of the comet, a much larger number of meteors is observed every day when the earth passes close to such an orbit. In fact, one then observes what is called a meteor shower, with all the meteors in the shower appearing to diverge from, or converge to, a single point in the sky. This convergence is only an apparent phenomenon, because the meteors are really traveling parallel to each other, so that they appear to be moving along lines that converge to or diverge from a point in the sky, since parallel lines appear to intersect at infinity. During the Leonid shower—one whose convergent point is in the constellation of Leo—of November 13, 1833, meteors were so abundant that as many as 200,000 per hour were seen at some observing stations. Such rich showers, however, are rare. Twelve distinct meteor showers, almost one a month, occur every year, but most of them are not very impressive. Each of these is associated with the orbit of a known comet.

If a large mass of either stony or metallic material enters the earth's atmosphere and if the collision is not too violent, it sometimes survives the vaporizing action of atmospheric friction to become a meteorite and strikes the earth with explosive force. If the impact is violent enough, a meteoric crater is formed and a good part of the meteorite disintegrates into tiny particles, which are dispersed in all directions. If the entire meteorite is not destroyed in this collision, the residue can be recovered for study. Sometimes one finds meteorites occurring in groups, as though they had entered the earth's atmosphere together or as though a single one had been broken into pieces during the collision. Since 1900 an average of six meteorites a year have been observed to enter the atmosphere and have been recovered, but obviously many more have remained undiscovered. The masses of meteorites range from a few grams to many tons.

There are characteristic surface features that enable one to recognize these objects almost immediately. The surface of the meteorite, if found soon after the fall, is very glossy and dark in color, indicating that it was probably formed from the fused material that remained while the meteorite was moving through the atmosphere. The surface itself is quite irregular and in some cases covered with deep pits, as though quantities of material had been vaporized out of it. Many of the meteorites are of stone; others are masses of alloys of iron, nickel, and cobalt. When meteorites are heated, they emit gases. About 75 of the known chemical elements have been found in meteorites.

Meteorites have recently become especially interesting because as many as 19 of the known amino acids that make up the proteins of living beings have been discovered on them. Since there is conclusive evidence that these amino acids were formed in interstellar space, the study of meteorites is very important in the understanding of the origin of life.

In addition to comets, meteors, and meteorites, which can be observed directly, there is a great deal of dust in the solar system that scatters sunlight, thereby producing a glow in the sky referred to as zodiacal light. Since the dust lies practically in the plane of the ecliptic, the zodiacal light itself can be seen under proper conditions in the western horizon after sunset as a luminous band extending upward from the horizon along the ecliptic. For an observer in the northern hemisphere the ecliptic is most nearly perpendicular to the horizon at sunset, when the sun is near the vernal equinox, so that he can best see the zodiacal light after sunset in the spring. The band of light is broad near the horizon but becomes much narrower as it approaches the Milky Way. The effect can also be seen in the east just before sunrise and is most pronounced during the fall.

Russian astronomers have recently suggested that zodiacal light is caused by a tail of dust, similar to a comet's tail, following the earth as it moves around the sun. Such a dust tail would also explain the interesting effect known as gegenschein, or counterglow—a faintly luminous diffused spot in the ecliptic that is about 180° away from the sun.

Early Theories of the Origin of the Planets

The pronounced regularity and symmetry of the solar system indicate a common origin for the sun and its planetary system; moreover, the near-circularity of the planetary orbits points strongly to an important connection between the rotation of the sun in its earliest stage, when it was contracting down gravitationally from a dust cloud to its present form, and the orbital motions of the planets. A remarkable and quite unexpected feature of the kinematics and the dynamics of the solar system is the way the total rotational motion in the solar system is distributed between the sun and the planets as a group. This rotational motion, or angular momentum, as it is called technically, for a single body of mass is expressed mathematically by a product of three terms: the mass of the body m, its speed v, and the radius r of its orbit, which may be assumed to be a circle. If the earth were actually moving at constant speed v in a circle of radius A around the sun, the product mvA would be the exact expression for its orbital rotational motion or angular momentum; but this is not true, so that this product is only a good approximation to the correct value. The discrepancy, however, is not important to this discussion. The important point is that the orbital angular momentum of a planet does not change as the planet revolves around the sun. This constancy of the angular momentum is known as the principle of the conservation of angular momentum. Since each planet has its own orbital angular momentum as it revolves around the sun, the total angular momentum of all the planets is obtained by adding together all the individual angular momentum terms. The numerical quantity obtained remains constant year after year. The relation of angular momentum to the rate at which a line from the sun to the earth sweeps out an area is shown in the diagram.

The total angular momentum of the planets may be compared with the angular momentum of the sun. Although the sun is actually moving in a very small orbit around a fixed point very close to its own center because of the gravitational

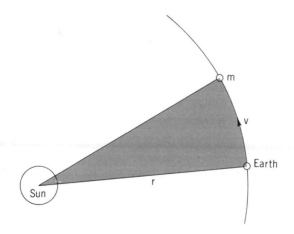

The rate at which the line from the sun to a planet sweeps out an area (the shaded area) as the planet moves in its orbit is a measure of the angular momentum of the planet. For a circular orbit of radius r this angular momentum is mvr.

pull of the planets—especially Jupiter—this motion contributes little to its angular momentum. Almost all the angular momentum of the sun stems from its rotation about its own axis. Since the sun rotates only once every 25 days, its total angular momentum is only a small fraction of the total angular momentum of the planets. This is a very puzzling state of affairs, for the sun contains more than 99.9 percent of the mass of the solar system, which would lead one to expect most of the solar system's angular momentum to stem from the rotational motion of the sun. The sun contains only 2 percent of the total angular momentum, so that 98 percent of the angular momentum in the solar system is concentrated in about 0.1 percent of its total mass. To see why this is so very peculiar and puzzling, note first that if the total mass of material in the solar system now is the same as it was in the original solar nebula, then owing to the principle of conservation of angular momentum, the total angular momentum present in the solar system now is the same as the amount that was present in the cloud of gas and dust from which the solar system was formed by gravitational contraction. The cloud was spread out over a distance of about 200,000 astronomical units and was rotating very slowly about an axis through its center but rotated more and more rapidly as it contracted. However, most of the rotational motion of the entire system must have remained in the contracting core, because the total angular momentum in a system cannot be altered by any changes within the system. Any of the outer regions of the cloud that were left behind when the core contracted could not have retained more than a small fraction of the total angular momentum. One is thus faced with the problem of explaining how the present distribution of angular momentum, with most of it outside the sun, was achieved. This could not have been the way it was initially, because if the outer regions of a cloud initially had as much angular momentum as the planets do now, the outer regions could never have condensed into planets at all.

It was precisely because of this problem with the distribution of angular momentum that such early theories of the origin of the solar system as the nebular hypothesis of the French astronomer and mathematician Pierre de Laplace (1749–1827) and the hypothesis of an encounter or collision between the sun and another star had to be discarded. Laplace's nebular hypothesis—which originated in the idea of the German metaphysician Immanuel Kant (1724–1804) that the matter now in the sun and the planets was originally spread out as a vast, rarefied cloud of gas—proposed that when the cloud contracted and rotated ever more rapidly rings of matter separated from it. Laplace argued that the centrifugal force at the periphery of the rotating nebular cloud became so large at a certain point of its contraction that a ring of matter was torn away and that this ring then contracted gravitationally to form the most distant planet first. This, according to the theory, was followed by the centrifugal breaking away of a second ring of matter as the central cloud contracted still further and rotated even faster; this second ring of matter was then supposed to have contracted to form the second most distant planet. And so it went, according to the Kant-Laplace nebular hypothesis, until all the planets were formed. Aside from the fact that such a simple breaking away of successive rings of matter from a centrally contracting body could not have led to the distribution of angular momentum in the present-day solar system, there are other fatal objections to the Laplace theory.

In the second half of the nineteenth century Maxwell demonstrated conclusively that even a ring of matter as massive as Jupiter spread around the sun could not have possibly contracted gravitationally into a spherical body; there would just not have been enough matter present in such a ring for this to have happened. Moreover, the contracting cloud could never have been rotating fast enough at any stage of its contraction to have thrown off rings of matter centrifugally. One can easily demonstrate this by calculating the speed of rotation that the matter at the sun's equator would have if all the planets were to fall into the sun; the angular momentum thus contributed to the sun by the planets would increase the sun's speed of rotation, but not nearly enough to cause rings of matter to break away. The centrifugal force at the sun's equator would still be only 5 percent of the inward gravitational pull. It is clear that if the centrifugal force would be too weak to tear rings away from the sun if all the rotational motion of the solar system were concentrated in the present sun, it would have been too weak to do so when the sun was greatly distended.

When Laplace's theory was discarded as the result of the above analysis, the collision hypothesis was proposed by astronomer Forest Ray Moulton (1872–1952) and geologist Thomas Chowder Chamberlin (1843–1928) in the United States and independently by Jeans in England. This theory assumed that an approaching star at some distant time in the past came very close to the sun and tore large chunks of matter out of the sun's surface that then fragmented and contracted gravitationally into the various planets. This collision theory and theories related to it, such as the tidal hypothesis proposed by Sir Harold Jeffreys, had to be abandoned also because they could not account for the large amount of angular momentum concentrated in the planets. Another serious and, indeed, unanswerable objection to the collision theory is that if matter were torn from the sun's interior, it would be so hot—at least 1 million degrees—that it could not even be held together by its own gravity, much less have contracted into spheres; the thermal motions of the molecules and atoms in such hot matter would be so violent that the atoms and molecules would disperse in all directions in a matter of seconds, like the fragments of an exploding bomb. In any case, a collision between the sun and another star in the Galaxy is so improbable that it cannot be considered seriously as a process that had anything to do with the formation of the planets.

Modern Theories

Since the big stumbling block in the way of Laplace's nebular hypothesis and the collision theories of the origin of the planets is the present distribution of angular momentum in the solar system, this problem was the first to be considered and, to some extent, solved in the most recent theories of the origin and formation of the planets, asteroids, and satellites. The first clue as to how the planets got their present disproportionate share of the total angular momentum in the solar system was pointed out for the first time in 1943 by the German physicist Carl von Weizsäcker: it can be found in the difference between the chemical composition of the four terrestrial planets as a group and that of the sun and the four major planets as a group. There are two distinct and sharp discontinuities in the chemical compositions of the terrestrial planets and the major planets. The sun is rich in hydrogen and helium (about 80 percent of the sun's mass is hydrogen, 17 percent is helium, and 3 percent consists of carbon, oxygen, and the metals); it has an average density of about 1.4 grams per cubic centimeter. The terrestrial planets consist of silicates, iron, magnesium, and other such heavy elements; they have average densities between 4 and 6 grams per cubic centimeter. The major planets are rich in hydrogen and helium; they have average densities between 0.75 and 1.6 grams per cubic centimeter and hence are chemically similar to the sun.

This sharp difference between the chemical properties of the terrestrial planets on one hand, and the sun and the outer planets, on the other, is a puzzle. One might have expected to find a gradual change in chemical composition of the planets and the sun or a gradual increase in density of the planets as one moves in toward the sun, just as one finds a gradual increase in the density of the earth's atmosphere as one moves from the top of the atmosphere in toward the earth. Weizsäcker used this very puzzling feature as the starting point and the basis of his theory of the origin of the planets. To explain the discrepancy in the compositions and densities of the two groups of planets and to account for the peculiar asymmetrical distribution of the solar system's angular momentum, Weizsäcker introduced the reasonable assumption that since originally the entire cloud of gas and dust from which the sun and planets were formed was chemically homogeneous, the flattened cloud of dust and gas from which the planets alone were formed was initially chemically homogeneous also, with the same hydrogen and helium abundances as in the protosun itself. Moreover, this disk of matter, according to Weizsäcker, was about 100 times more massive than the present total mass of the planetary system, with most of it in the form of gaseous hydrogen and helium.

Accepting this picture, one can now explain how the present, unusual distribution of angular momentum in the solar system occurred and why the chemistry and density of the inner planets are different from those of the outer planets. First, it should be noted that owing to the gaseous viscosity of the disk, the protosun and the disk would tend to rotate like a rigid body; the gaseous friction between the protosun and the disk tended to keep the disk rotating in step with the protosun. This means that the disk was set rotating at very high speeds—some thousands of kilometers per second—at the expense of the protosun and thus had most of the angular momentum of the system then just as it does now. There is thus nothing to explain, since what one sees now is a natural consequence of the dynamics of the initial system, in which the gaseous viscosity transferred angular momentum from the contracting core to the disk. As the speed of rotation of the contracting protosun increased, the protosun dragged the outer disk with it. In this way, angular momentum was constantly transferred from the protosun to the planetary disk.

To account for the difference in the present chemical compositions of the inner and outer planets and also for the much smaller mass in the present disk than in the initial disk, three things must be considered: first, the difference between the temperature of the inner disk and the temperature of the outer disk; second, the difference between the masses and chemical affinities of the hydrogen and helium atoms and the masses and chemical affinities of the heavier atoms like carbon, magnesium, and silicon; third, the solar wind and the solar magnetic field. As the protosun contracted and became hot enough to radiate energy, the temperature of the rotating disk increased as it absorbed energy, but the inner disk became much hotter than the outer disk. Owing to this, the atoms and molecules of the gaseous material in the inner disk acquired much higher speeds than did these same kinds of atoms in the outer disk. Indeed, the speeds of these light atoms in the inner disk were high enough for most of these atoms to escape from the inner disk, and this escape was aided by the pressure of the solar radiation and the stream of solar particles. Moreover, the high temperature in the inner part of the disk produced ionization of the hydrogen and helium atoms, with the result that these ions were accelerated outward by the lines of force of the solar magnetic field. The heavy elements in the inner disk were not affected by these forces for two reasons: first, owing to their larger masses the speeds of these atoms were not high enough for them to escape from the inner disk; second, owing to their chemical affinities these heavy elements clumped together to form grains of dust that, because of their large masses, moved about very sluggishly and thus had very little chance of escaping; the grains of dust were also massive enough for their gravitational attraction to the protosun to withstand the outward drive of the solar wind. The grains of dust had another important effect on the inner disk; by effectively absorbing the solar radiation the dust particles caused the temperature of the inner disk to increase considerably more than it otherwise would have and thus contributed to the escape of hydrogen and helium atoms. In this way, the inner regions of the nebular disk were fairly rapidly depleted of the light gaseous elements hydrogen and helium, leaving the field to the relatively small quantities of the heavy elements, which now constitute more than 99 percent of the matter in the inner planets.

One other effect—arising from the magnetic field of the initial sun—that probably contributed considerably to the transfer of angular momentum and mass from the protosun and the inner disk to the outer regions was the acceleration of ions by the solar magnetic lines of force. Evidence collected by various space probes indicates that a variety of disturbances on the sun's surface today eject high energy ionized particles —mostly protons and some alpha particles—into space, giving rise to the solar wind. The ionized particles in this wind follow the sun's magnetic lines of force, which form large loops that begin and end on the sun's surface but extend far out into space. As the sun rotates, these magnetic loops rotate with the sun, dragging the ionized particles with them as though these particles were frozen to the magnetic field. The effect of this is to slow down the rotation of the sun itself and to speed up the rotation of the outer regions of the disk. Although this effect is small today, it was probably very large when the sun was just contracting down to its present size because of the high density of ionized particles surrounding the sun then and the large magnetic field intensities. When the sun was first formed from the contracting cloud, both it and its magnetic field were probably spinning much more rapidly than they are now and much more rapidly than the disk was spinning. Keeping in mind that the magnetic lines of force behaved like rigid cables whose ends were attached to the sun, one sees that the ions moving along these lines of force were set rotating, with the result that the nebular disk itself was dragged around and set spinning more and more rapidly but the rotation of the contracting sun was slowed down. All of these effects, which resulted in the transfer of angular momentum and mass from the protosun and the inner disk to the outer disk, eliminate the difficulties with the distribution of angular momentum that led to the rejection of Laplace's original nebular hypothesis. That is why the modern theories of the formation of the planets are all variations of Laplace's original theory with some very important changes.

Note that the effects that drove the light elements out of the inner part of the nebular disk and into the outer regions of the disk were very weak at great distances from the sun. At a distance of 500 million miles from the sun, the mean distance of Jupiter, the intensity of the solar wind and the intensity of the solar radiation were probably about one twenty-fifth of what they were in the neighborhood of the earth. That is why the hydrogen and helium in these outer regions were not blown entirely out of the solar system, even though large quantities were. To be sure, the speed of escape from the sun at these large distances is much smaller than near the earth, but offsetting this is the very low temperature at these distances and therefore the very small molecular speeds, which favored the accretion of molecules and led to the formation of the massive planets. The formation of these massive planets occurred rapidly enough so that their large gravitational fields were able to retain enough of the light elements to give them their present chemical compositions in which hydrogen and helium are, by far, the most abundant components. Even so the outer disk also lost a great deal of its initial mass in the form of outward-flowing hydrogen and helium.

8 *How the Earth Was Born*

Creation of the Planets

Even before the dynamical effects previously described brought about the present distribution of mass and angular momentum, the planets were well on their way to being formed; indeed, it is highly likely that the planets were formed quite rapidly even before the protosun had contracted down to its present form. The first step in the formation of the planets occurred when the solar nebula broke up into a hierarchy of turbulences and eddies of various sizes. It can be shown that a rotating disk having a mass equal to 100 times the present total mass of the planets and rotating as fast as the planets are now rotating would have broken up into all kinds of turbulences because of the differential motion between various parts of the solar nebula; the Reynolds number for such a rotating disk would greatly exceed the critical value of 1,000, which marks the onset of turbulence. These turbulences, consisting of vortices and eddies of various sizes, were the first step in the formation of the planets, because they led to the accretion of dust particles into large lumps of matter at varying distances from the protosun; an eddy, or vortex, is a dynamical structure that consists of swirling particles of matter and that maintains itself against dissipation by means of its own angular momentum. As the eddy moves along, it tends to collect more and more matter, which leads to a constant accretion if the conditions are appropriate. Since hydrogen molecules and helium atoms cannot condense into grains of dust even at temperatures close to absolute zero, the accretion that occurred in the turbulences in the solar nebula involved only the heavy elements that had condensed into dust grains. But only certain of the turbulences that arose led to further accretions of dust grains into planetlike structures because most of the turbulences were quickly disrupted and dispersed by the tidal action of the protosun.

Just as the moon raises tides on the earth, causing the earth to stretch slightly along the line from the center of the earth to the center of the moon, so did the protosun cause each turbulent eddy and accretion of matter to stretch along a line from the eddy toward the protosun. Now, it can be shown that this kind of tidal action disrupted and destroyed any given eddy or condensation of dust grains unless the density of such a condensation exceeded a certain critical value that depended only on the mass of the protosun and on the distance of the condensation from the protosun. The distance from the protosun at which an accretion of matter of a given density was disrupted by the tidal action of the protosun is known as the Roche limit for that density; the larger the density, the lower the Roche limit (that is, the closer to the protosun could such an accretion occur without being disrupted). The densities of the planets, starting with Mercury, are just what one would expect to find if these planets arose out of turbulences and accretions that were formed just beyond the Roche limits for these densities. Since Mercury is closer to the sun than any other planet, one would expect its density to be larger than that of any other planet, which is, indeed, the case. Turbulences and accretions of dust at Mercury's distance but with densities less than Mercury's could not have withstood the tidal disruption of the protosun. Here, then, is a reasonable explanation of the relationship between the mean densities of the planets and their mean distances from the sun. The formation of the sun and the four inner planets is illustrated in the diagram.

When the planets were formed, it is probable that the sun had not yet become hot enough to radiate much energy, so

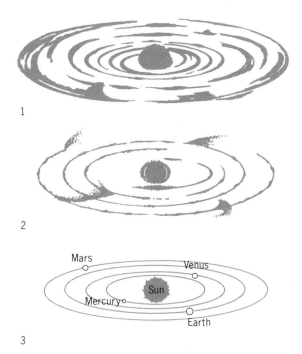

The three stages of the formation of the sun and the four inner planets from a gas cloud. The turbulent motion in the cloud surrounding the protosun and the tidal action of the protosun give rise to condensations at definite distances from the center of the protosun.

that darkness prevailed at that time. Owing to this, the initial stages of the planets, or protoplanets, consisted of solid cores surrounded by extensive atmospheres consisting largely of hydrogen and helium. But this situation did not last very long: once thermonuclear reactions were set going in the sun, the intense radiation from the sun and the stream of solar ionized particles swept away most of the hydrogen and helium in the atmosphere of the planets. In the case of the four inner planets, all of the free hydrogen and helium atoms were driven off, leaving these planets pretty much as they are now. In the case of the four outer planets, the solar radiation and solar wind were not strong enough to do a complete job; a good deal of the hydrogen and helium was torn away from these massive planets but enough was chained gravitationally to them to give them their present chemical composition.

ROLES OF TEMPERATURE AND CHEMISTRY

Although certain objections have been raised to these theories of planetary formation and certain questions are still unanswered, there appear to be no other primary agents that could have brought about the present planetary distribution of mass, chemical composition, and angular momentum. However, there were other, secondary influences at work in the formation of the planets that must be considered briefly if one is to understand some of the finer details observed in the properties of the planets. The most important of these agents is the temperature, which greatly influences the chemical equilibrium in a gaseous mixture of different chemical elements and determines the rate of condensation of gaseous matter into solids. The American geophysicist Harrison Brown, a pioneer in chemical cosmogony, pointed out in 1950 that there are three major groups of chemical compounds in the solar system that solidify at different temperatures: the metallic oxides and rocky materials, such as the silicates, aluminates, and titanates; the icy materials, such as water, ammonia, and methane; and the gases such as hydrogen, helium, and neon.

Consider a mixture similar to the initial solar nebula of all of these substances at such a high temperature that they are all in a gaseous state, as would be the case if the temperature of the mixture were higher than 2,200°K. A state of chemical equilibrium exists among these various gases under these conditions, but this state of equilibrium changes as the temperature changes. As the temperature drops, some of the gaseous components condense out of the gaseous phase to become liquids and solids. Thus, below a temperature of 1,500°K but above 1,000°K, such metallic compounds as calcium titanate, magnesium silicate, iron, and the alkalisilicates and aluminosilicates condense out as solid particles. These are just the kinds of compounds that are found in the rocks and metallic ores that constitute the crust of the earth, moon, and probably those of the other inner planets. Above a temperature of 1,200°K the sodium and potassium atoms are still in a gaseous state, but below this temperature they combine with aluminum, silicon, and oxygen to form solid particles of the aluminosilicates. Below 1,000°K but above 500°K the condensed metallic iron combines with sulfur to form iron sulfide, which is found in all the meteorites; at the same time, the hydrous, or water-bearing, calcium silicates are formed and the ferrous oxide molecules, which had condensed out previously, combine with magnesium and the silicon oxides to form the ferromagnesium silicates. At a slightly lower temperature the very abundant mineral talc is formed. If one applies this analysis of the first of the three major condensations from a gaseous mixture as the temperature drops to the initial chemical mix in the solar nebula, one can see how the present chemical composition of the inner planets came to be.

As the temperature of the gaseous mixture continues to drop, the various ices, such as water, ammonia, and hydrated methane, condense out to form the kinds of materials that constitute the crusts of the Jovian planets. The condensation of the ices begins at a temperature of about 150°K when methane solidifies. Below this temperature the noble gas argon solidifies, but hydrogen and helium remain gaseous down to temperatures very close to absolute zero, so that the condensation of these two lightest gases does not occur on the surfaces of any of the planets or satellites in the solar system. The condensation of these gases is therefore not of much interest as far as the solar system is concerned.

The manner in which the various chemical components in a hot gas condense out as the temperature of the gas falls was used by Harold Clayton Urey, American Nobel laureate in chemistry, to develop his chemical theory of the formation of the planets, which, together with the theory of turbulences described in the previous section, can account for most of the physical, dynamical, and chemical properties of the planets. According to Urey's analysis, the planets were all formed at relatively low temperatures—below 1,400°K—by the accretion of the kinds of condensed particles described above; this accretion aided by turbulences led to the formation of planetesimals of various sizes and masses revolving around the sun in various orbits. At any particular distance from the sun, only those planetesimals survived that were dense enough to withstand the tidal disruption of the sun at that distance. This ultimately was determined by the kinds of elements in the solar nebula that could condense into solid particles at the given distances; it was shown above that these were just the silicates and the metals, which compose the inner planets. Here, then, we have an explanation of the distribution of chemical elements among the various planets based on the theory of chemical equilibrium.

After planetesimals of various sizes were formed by condensation from the solar nebula, they collected, with the aid of turbulences and accretion, to form the planets. Initially, accretion occurred entirely as the result of random collision between the planetesimals of various sizes. When two planetesimals of equal mass collided, they tended to pulverize each other, but a planetesimal of small mass tended to bury itself in a larger one and merge with it on collision. Thus, random collisions between planetesimals in this early stage of planetary formation led to the gradual disappearance of smaller particles and their amalgamation into larger ones. In time a few of these large bodies became massive enough through such collisions to attract to them gravitationally most of the particles that were revolving around the sun. This greatly accelerated the accretion process and finally led to the formation of the planets pretty much as they are now.

It is clear that the planets must originally have been chemically homogeneous because the planetesimals themselves were. But in time separation between the dense iron-nickel components and the silicates occurred. The reason for this was that the interiors of the planets were heated by the radioactivity of the uranium ores there and the constant bombardment of the surfaces of the planets by planetesimals. This heating caused the interiors of the inner planets to melt, with the result that the iron-nickel components of the planets sank to the center to form the dense cores. The silicates and various uranium ores rose in the form of gases to become the outer mantle. It should be noted that direct evidence for this separation of the silicates and the iron-nickel is found in the different kinds of meteorites that regularly strike the earth's surface. Two such kinds of meteorites have been found: the rocklike meteorites and the metallic meteorites. The former consist of the silicates (mostly silicon dioxide or sand), magnesium, some iron, some calcium, and heavy elements like uranium. The metallic meteorites consist of iron (90 percent), nickel (8.6 percent), and small amounts of cobalt. It appears, then, that the same forces that separated the iron-nickel components from the silicates in the earth produced the two kinds of meteorites, so that the existence of these two kinds of meteorites strongly favors the above description of the way the dense cores of the inner planets separated out of the molten interiors of the inner planets.

CREATION OF THE EARTH

The formation of the earth is of special interest because, as far as is known, it is the only planet in the solar system that supports life. The earth, like the other planets, was formed by the gravitational accretion of planetesimals that were revolving around the sun at the same mean distance as the earth is now, but the structure of the earth differs from that of Mercury and Venus in certain important details that can be explained if one takes into account the differences in the original physical conditions that surrounded Mercury and Venus when these planets were formed. Mercury was the hottest of the three planets, being closest to the sun, so that it was formed at a temperature so high that the silicates had not yet condensed. Owing to this, Mercury developed the largest iron core and is thus the densest of the three planets. The temperature of Mercury was much too high for any water to condense—a very important difference between Mercury and the earth. The difference in temperature between the earth and Venus when they were formed was not very great, but it was sufficient to account for the absence of both sulfur and water on Venus and their presence on the earth. The earth coalesced in a region in which iron sulfide and the hydrous silicates condensed. Thus, whereas Venus contains hardly any sulfur or water—if the analysis based on chemical equilibrium is correct—the earth has a great deal of sulfur and some water.

The presence of sulfur in the earth when it was formed played an important role in its thermal history, which was quite different from that of Venus. The mixture of iron and iron sulfide that permeated the earth when it was formed as a cold body by accretion of planetesimals had a fairly low melting point as compared to the melting points of the materials inside Venus. Owing to this, the iron–iron-sulfide mixture in the earth liquefied quite early in the geological history of the earth as it warmed up owing to the radioactivity of its uranium. This probably happened about 4 billion years ago or even earlier, so that the interior of the earth probably began to melt even before accretion of planetesimals ended. This had two important consequences: First, the earth's iron-nickel core and, hence, the earth's magnetic field were formed some 3.5 billion years ago. Second, the differentiation of the earth into core, mantle, and crust occurred quite catastrophically, somewhat like a gravitational collapse, with the gravitational release of large quantities of heat traveling like a pulse to the surface and setting off volcanoes and earthquakes, which had a very important influence on the formation and evolution of the earth's atmosphere.

ORIGIN OF THE EARTH'S ATMOSPHERE

In considering the atmospheres of the planets, with particular emphasis on the earth's atmosphere, one first notes the difference between the atmospheres of the four terrestrial planets and those of the four Jovian planets, which, although remarkable, is not too difficult to explain in terms of the general picture of the formation of the planets painted above. The chemical nature of the dense atmospheres of Jupiter, Saturn, Uranus, and Neptune was determined by the great abundances of hydrogen and helium in the original solar nebula. Since the Jovian planets are much more massive than the earth (even the least massive of these, Uranus, has 14.5 times as much mass as the earth), the original hydrogen and helium that surrounded these planets could not escape from them, because of their large gravitational fields. The very large hydrogen abundances led to the formation of molecular hydrogen, methane, ammonia, and water in these atmospheres with the result that no free carbon, nitrogen, and oxygen are now present in these atmospheres; the dominant components are molecular hydrogen and atomic helium. An atmosphere of this sort is called a reducing, or hydrogen-rich, atmosphere which distinguishes it from the earth's atmosphere, which is an oxidizing, or oxygen-rich, atmosphere.

Whereas the present atmospheres of the Jovian planets are the same now as they were originally, the atmospheres of the terrestrial planets evolved to their present chemical composition in two stages. During the first stage, the atmospheres of the inner planets, called the primitive atmospheres, were similar to those of the Jovian planets, but these primitive, reducing atmospheres did not last very long. During the second stage of atmospheric development, the terrestrial planets lost their primitive atmospheres; and except for Mercury, which has no atmosphere, each of these planets acquired from its interior, through volcanic action, a secondary atmosphere, which in the case of the earth is highly oxidizing.

The most conclusive evidence that the terrestrial planets lost their primitive atmospheres shortly after the planets were formed is found in the composition of the earth's atmosphere itself. If the earth's present atmosphere were the same as its primitive atmosphere, it would now have the same composition as the atmospheres of the Jovian planets. It is not surprising that there is no free hydrogen or helium in the earth's atmosphere; the atoms of these gases are so light that they escaped not only from the earth but also from Venus and Mars. But what is surprising is that the heavy noble gases neon, argon, and xenon are present in the earth's atmosphere in only very minute quantities. Since neon is one of the most abundant elements in the universe—it is more abundant than nitrogen—one would expect to find far larger quantities of it in earth's atmosphere than there actually are if something had not happened to the earth's primitive atmosphere. The actual abundance of the two other noble gases, argon and xenon, in the atmosphere is also far smaller (about one-millionth as large) than what one would expect. Since these gases are chemically inert, their practical absence from the earth's atmosphere could not have been caused by their reaction with other elements to form solid compounds that precipitated out of the atmosphere. The only explanation of their extreme scarcity in earth's atmosphere is that, following the escape of hydrogen and helium, they were swept away into outer space with the rest of the atmosphere, so that at the end of the first phase the earth, like Venus and Mars, was circling the sun as a bare, atmosphereless, rocky sphere.

Where and how, then, did the earth get its present oxidizing atmosphere which is so rich in nitrogen and oxygen, but almost entirely free of carbon dioxide? These gases could only have come from the earth's interior as the result of volcanic activity. When the interior of the earth melted and the iron-nickel component sank through the molten mass to form the dense core, intense volcanic activity was produced. This volcanic activity released large quantities of water vapor, nitrogen, and carbon dioxide. One can show from a study of the chemical composition of the gases released from volcanoes at the present time that volcanic activity on the earth during the last 4 or 5 billion years can easily account for the present abundance of nitrogen in the atmosphere and for the total amount of water vapor that was required to fill all the oceans with the amount of water they now have.

But what about all the carbon dioxide that was also emitted during all this volcanic activity? Where did it all go, for 10 percent of the gases emitted by volcanoes now is carbon dioxide? One must answer this question, for it can be shown that if the rate at which carbon dioxide was emitted during volcanic activity in the past was at least as large as it is now, and there can be no doubt about this, the amount of carbon dioxide in the atmosphere now should be about 100 times as great as the amount of nitrogen. If the earth had not rid itself of its carbon dioxide in some way in the past, its atmosphere now would consist of more than 99 percent carbon dioxide, and the atmospheric pressure would be no less than a half a ton per square inch instead of the 15 pounds per square inch that it is now. Moreover, the temperature at the surface of the earth would be very high. Fortunately for man, however, and for life in general on the earth, the carbon dioxide was removed from the earth's secondary atmosphere by a very simple chemical reaction involving the calcium silicate in the surface rocks of the earth. In this reaction the gaseous carbon dioxide combines very readily with calcium silicate to form quartz, which is to say, all the sand on the earth, and calcium carbonate, which appears on the earth in the form of limestone and marble.

This removal of most of the carbon dioxide from the

earth's secondary atmosphere was extremely important for the origin of life on this planet. Carbon dioxide molecules, which give rise to a greenhouse effect, allow the visible rays of light from the sun to pass quite readily through the atmosphere and to heat up the surface of the planet, but they do not allow the infrared rays and heat rays emitted by the warm surface to leave the planet, so that the surface of the planet gets very hot in time if the atmosphere contains enough carbon dioxide. Indeed, as happened to Venus, which could not get rid of its atmospheric carbon dioxide (the atmosphere of Venus, as shown by the Russian Venus probe, is more than 95 percent carbon dioxide), the surface temperature of the earth would now be almost 800°F. Thus, life would never have originated on earth if the earth's atmosphere had not lost its original heavy concentration of carbon dioxide. The abundance of carbon dioxide in the earth's atmosphere now is so small— only 0.03 percent—that it has very little effect on the temperature of the earth's surface.

After the calcium silicate reaction described above had reduced the abundance of the earth's atmospheric carbon dioxide to its present small value, the earth's surface temperature became cool enough for life to begin and to flourish, and this early life probably helped maintain the carbon dioxide balance when the earth was still very young. The very earliest marine animals absorbed the carbon dioxide molecules dissolved in the oceans, which they then converted into their own shells as solid calcium carbonate, but this did not deplete the carbon dioxide in the oceans because for each molecule of carbon dioxide that was taken in this way from the oceans, another was absorbed by the oceans from the atmosphere. This did not decrease the abundance of carbon dioxide in the atmosphere, because volcanic action, although greatly reduced, continued to expel carbon dioxide from the earth's interior in sufficient quantity to balance the amount of it absorbed by the oceans. Thus, the carbon dioxide balance was maintained.

Although the rocks on Venus contain the same kind of calcium silicates and in the same abundance as do the rocks on the earth, the surface of Venus was never cool enough for the calcium silicate in its rocks to remove much carbon dioxide from its atmosphere. The chemical reaction that very effectively removed carbon dioxide from the earth's atmosphere early in its history is very sensitive to the temperature of the rocks. The higher the temperature, the more slowly it operates. Since Venus is about one third closer to the sun than the earth, its surface was considerably hotter than the earth's surface was when the planets began to acquire their secondary atmospheres. Because of this the rocks on Venus absorbed very little carbon dioxide from its atmosphere and a greenhouse effect ensued; this raised the surface temperature of Venus still higher, with the result that even more carbon dioxide accumulated in its atmosphere and the greenhouse effect became stronger. In time enough carbon dioxide had collected in Venus' atmosphere to bring the atmosphere to its present chemical composition and to raise the temperature of Venus to its present high value of about 800°F.

One more point to clear up about the earth's atmosphere is the large abundance of free oxygen. If this element were not being constantly resupplied to earth's atmosphere by some process, it would quickly disappear from the atmosphere because it combines readily with so many things at normal temperatures to form solid oxides. It is clear from this that the oxygen in the atmosphere was not present when the earth acquired its secondary atmosphere but accumulated slowly as the result of plant life. Plants require very little oxygen for their life processes, but they convert great quantities of carbon dioxide into free oxygen, which they release to the atmosphere, and into carbon, which they build into their own organic structures. The presence of free oxygen in the atmosphere in turn plays an important role in maintaining the water balance on the earth. When three oxygen atoms combine, which happens quite frequently about 50 miles above the earth's surface, a molecule of ozone is formed, and at the present time enough such ozone molecules exist to form a layer of ozone above our atmosphere that protects man from the sun's dangerous ultraviolet radiation. Ozone readily absorbs ultraviolet rays, so that the intensity of the ultraviolet radiation from the sun is greatly reduced and thus rendered harmless after it has passed through the ozone layer. If this did not happen, the earth would, in time, lose its water because the intense ultraviolet radiation would break up the water molecules into hydrogen and oxygen and the hydrogen would then escape into outer space.

The marvelous chemical balance of carbon dioxide, oxygen, and plant life never got started on Venus because Venus could not get rid of most of its carbon dioxide, which produces a runaway greenhouse effect. This, in turn, produced a very high surface temperature, which prevented plant life from developing there; without plant life Venus could not accumulate any free oxygen in its atmosphere. Finally, without any free oxygen to shield its water-vapor molecules from dissociation by solar ultraviolet radiation, Venus lost most of its water. Thus, instead of being a pleasant, life-supporting planet, Venus is a hot, arid planet with a very dense carbon dioxide atmosphere. The vast difference between conditions on the surface of the earth and on Venus arose because Venus began its life as a planet a little bit too close to the sun; if it had been born a few million miles farther from the sun than it is now, the same calcium silicate–carbon dioxide chemical reaction that removed most of the carbon dioxide from the earth's atmosphere would have done the same thing for the Venusian atmosphere, and life would have taken hold and evolved on Venus just as it did on the earth. The earth just missed having a real sister planet, which would have been a magnificent biological laboratory for man—or perhaps vice versa.

It is clear that planets had to lie a certain minimum distance from the sun if they were to escape a runaway greenhouse effect and develop an atmosphere that could support life. The mean distance of the earth from the sun seems to be just the right distance for a planet for the evolution of an incredibly rich variety of life forms, but this does not mean that this is the only permissible distance for life. The thermal history of Venus tells us how close to the sun a planet can be before the origin of life becomes impossible, but it is still not known how far away from the sun a planet can be and still sustain life. To answer this question, one can turn to Mars, but the answer from Mars will not be conclusive, since many different factors, in addition to the radiation from the sun, are involved. All the evidence obtained from recent Martian space probes indicates that Mars has a very tenuous atmosphere which contains a high percentage of carbon dioxide but very little free oxygen, nitrogen, or water vapor. The present Martian atmosphere is undoubtedly a secondary atmosphere, but it has been greatly influenced by the lack of volcanic activity on Mars. Since the mass of Mars is only about 11 percent that of earth, its internal temperature was, and is, much lower than the earth's, so that volcanoes on Mars were rare and of little importance. For this reason, very little nitrogen or water vapor accumulated on Mars, and so, the Martian atmosphere remained very thin. The small mass of Mars also was responsible for its small gravitational field, which allowed the light gases to escape. Owing to this, the greenhouse effect on Mars was quite ineffective in keeping the surface of Mars warm; its mean surface temperature today is about $-40°F$, much too low to support life. This does not mean that life would not have developed on Mars if it had been as massive as the earth; chances are that it would have, for its secondary atmosphere would then have been as dense as, and chemically similar to, earth's. Such an atmosphere would have given Mars a nice warm blanket under which life would have arisen.

Age of the Earth and the Solar System

Meteorites and comets permit one to obtain direct evidence of the conditions in the outer regions of the solar system and to draw some conclusions about the state of the solar system at its birth. In addition, these objects can be used to check estimates of the age of the earth and the solar system that are based on an analysis of the earth. Meteorites and comets are probably among the oldest objects in the solar system because they have undergone very little change since they were formed from the primordial solar nebula. It appears that meteorites were formed some 4.5–5 billion years ago in the hot interior of some parent body. The material from which the stony meteorites were formed rose to the surface of this molten body and the heavy material from which the metallic meteorites were formed sank to the core. When the parent body ultimately disintegrated, which may have happened when the asteroids were formed—indeed, the meteorites may have been formed from the same source as the asteroids—the stony material from the surface and the metallic material from the core were separated and condensed to form the two kinds of meteorites now observed in the solar system. From the remnants of meteorites that strike the earth's surface one can deduce the age of the meteorites and, hence, of the solar system and obtain a good deal of information about conditions in outer space.

Various methods can be used to deduce the age of the solar system from the analysis of meteorites, but the most common one, until recently, was to determine the amount of helium retained by the meteorite after it had been formed from its parent body and had cooled. While the meteorite was hot, most of the helium escaped, but when it froze, helium began to accumulate in its interior because of the radioactivity of uranium. The two natural isotopes of uranium, uranium-238 and -235, decay spontaneously into other radioactive elements, finally ending up as isotopes of lead. During this process helium is emitted and accumulates inside the frozen meteorite, remaining trapped there until the meteorite strikes the earth. Knowing the rate at which uranium atoms decay, one can then determine the age of the meteorite by measuring the concentration of helium, relative to that of uranium, in the meteorite. This procedure gives an age of about 4.5 billion years for meteorites and, hence, for the parent body and the solar system itself. This agrees well with estimates of the age of the earth based upon radioactive-isotope dating techniques applied to rocks on the earth. However, the helium dating technique for determining the ages of meteorites is somewhat uncertain because of the production of helium in meteorites by the bombardment of meteorites by very energetic cosmic rays, which originated in such events as supernovas and which permeate all of space and move about in all directions; they consist of very high speed particles (mainly protons, traveling at speeds up to 99.99 percent of the speed of light) that disrupt the nuclei of any atoms that they strike. Owing to such collisions cosmic rays transform some of the nuclei in meteorites into other nuclei and leave a residue of helium nuclei in the meteorite. This so-called cosmogenic helium, which can be quite abundant and which might have been formed at any time after the birth of the meteorite, introduces errors in the estimate of the age of meteorites based on the measurement of the relative abundances of helium and uranium nuclei in the meteorites when they hit the earth.

But the very cosmic rays that lead to errors in the estimate of the ages of meteorites based on the helium-uranium ratio leave their marks on meteorites, enabling one to determine how long these objects have been exposed to cosmic rays and, therefore, how old they are. The energetic particles in cosmic rays transform some of the surface nuclei on meteorites into new nuclei, some of which are stable and some radioactive; this process is called spallation. By measuring the half-lives and the abundances of these nuclei, one can obtain the length of time the meteorite had been wandering around before it hit the earth and, hence, the age of the meteorite. The results thus obtained agree with the estimates of the age of the solar system obtained from lunar rocks and soil and from the uranium-lead ratio in terrestrial rocks. Not only do cosmic rays transform the nuclei in meteorites into other nuclei, but they actually leave tracks on the surfaces of meteorites that can be photographed with electron microscopes. These tracks contain a fund of information about the nature of cosmic rays, the conditions in space where the meteorites were formed, and even the size of the parent body from which meteorites were formed.

In recent years meteorites have come under more intensive scrutiny by scientists than ever before because a number of different organic compounds have been found on their surfaces, but what is more important is that among these organic compounds 19 of the 20 known amino acids have been identified. The importance of this for the whole question of the origin of life is obvious.

Although one cannot get hold of a comet as easily as one can of a meteorite, there are circumstances in the life of a comet that force it to reveal a great deal about itself. When a comet's orbit is very elongated, it passes close to the sun along part of its orbit and the radiation from the sun then excites the atoms of the comet. From the spectrum of the radiation emitted by these excited atoms, one can determine the chemical constitution of the comet; such a spectral analysis reveals that comets consist primarily of ices and snows of various kinds. The dominant constituent of comets is ordinary ice, but solid carbon dioxide, solid methane, and other compounds of the light elements have been identified. In the case of the recent comet Kohoutek, the organic compound methyl cyanide was identified as one of its constituents. The chemical composition of the comets indicates that they were formed from the same primordial material as the outer planets. The comets probably condensed from the extremely cold outer fringes of the solar nebula very early in the history of the solar system, and because of the very low temperature in these outer regions, hydrogen and helium were probably also captured in the interior of the frozen comet. The comets must thus be pictured as large masses of ice and snow circling in vast orbits around the sun; they are the residue of the solar nebula, left behind when the main body of this nebula contracted to form the sun and the planets.

The age of the meteorites gives the age of the solar system, but the methods for determining meteorite ages are still somewhat unreliable, so one must rely on the procedures for determining the earth's age to obtain an accurate value for the age of the solar system. There are various geological and physical methods for doing this, among which is the measurement of the present saltiness of the ocean. When the surface of the earth cooled below the boiling point of water some hundreds of thousands of years after it had formed as a solid crust, there were no oceans; all the water that had been ejected from the interior by volcanic action was present in the form of water vapor as one of the constituents of the earth's atmosphere. But then torrential rains filled the deep valleys on the cool earth's crust with large bodies of fresh water, which were the original oceans. The present salinity of the oceans is the result of a slow accumulation that went on for a few billion years; it was the work of the rivers that carried, and still carry, millions of tons of all kinds of salts into the oceans every year. Knowing the present salinity of the oceans as well as the rate at which all the rivers are transporting salt to the oceans, one can calculate how long it took the oceans to become as salty as they are now. Although there are some uncertainties in this kind of calculation the method is a reasonable one and gives a figure of a few billion years for the age of the earth.

A procedure for estimating the age of the earth that is closely related to the salinity-of-the-oceans method is that of measuring the thickness of sedimentary rocks in regions that were once under some ocean but are now exposed dry land, such as the Grand Canyon of the Colorado. Sedimentary rock is formed at the bottoms of the oceans by the pressure of the ocean water and the upper layers of sediment acting on still lower layers of silt, which was transported to the ocean by the rivers in the past. The rate at which sedimentary rock is built up from this silt can be calculated from the amount of silt that is deposited in the oceans by rivers every year. Knowing this and the total thickness of the sedimentary rocks on the earth, one can calculate how long it took these rocks to be formed by sedimentation and pressure. The figure obtained is again about 4.5 billion years, which agrees with the age of meteorites and the age of the oceans.

The most accurate and reliable method for determining the age of the earth is based on measurements of the ratio of the abundance of uranium to that of lead in the crust of the earth. The nuclei of uranium and thorium present in small amounts in the rocks that form the crust of the earth are radioactive; they decay slowly into stable lead nuclei. By comparing the number of stable lead nuclei that originated from the radioactive ones with the number of radioactive ones still left in a given specimen of rock, one can find the age of that rock, since the rate of decay of the radioactive nuclei is known. Direct measurements of the radioactivity of a given quantity of ordinary uranium have shown that half of the nuclei in this quantity decay into stable lead nuclei every 4.5 billion years; this is called the half-life of uranium-238. Once the crust of the earth hardened, which probably happened a few million years, at most, after the earth was formed, lead derived from the decay of uranium began to accumulate in uranium ores, so that the amount of lead in uranium ores permits an accurate estimate of the age of the earth and, hence, of the solar system. The age obtained in this way is again about 4.5 billion years.

9   *The Structure of Atoms and Molecules*

A Hierarchy of Microscopic Structures

Each member of the remarkable hierarchy of gravitational structures in the universe in turn consists of particles of matter that can also be arranged in a hierarchy of structures that are governed by nuclear and electromagnetic forces. In the deep interiors of stars where the temperatures are measured in tens of millions of degrees, there are no intact atoms but only a hierarchy of nuclear structures, ranging from single protons through alpha particles and carbon nuclei up to iron nuclei and even to those beyond uranium in the periodic table of chemical elements. In the deep interiors of stars the temperatures are so high that only the nuclear force is strong enough to bring some order out of the violent, chaotic motions of the constituent particles. This is a subatomic order, as exhibited by the existence of various nuclei of increasing mass and complexity. A very powerful electron microscope reveals that the neutrons and protons inside any given nucleus are arranged in a system of shells somewhat similar to the "electron shells" inside an atom. Thus, even within such tiny particles as nuclei, with diameters of the order of a few tenths of a trillionth of a centimeter, order prevails under the influence of the strong nuclear forces. In the deep stellar regions the electromagnetic forces, although much stronger than the gravitational force, are much too weak to construct complete atoms by uniting the rapidly moving nuclei and electrons that dart about in all directions. To be sure, at any given moment some of the randomly moving, negatively charged electrons will be captured by some of the positively charged nuclei to form ions momentarily, but these ions will be quickly broken up again into free nuclei and electrons by collisions and by the intense gamma radiation streaming out from the star's center, while other nuclei and electrons recombine to form ions. Thus, the most the electromagnetic force can do deep inside stars is to interrupt the wild motions of groups of nuclei and electrons momentarily to form temporary, partial atomic structures. One cannot speak of any real atomic order in these stellar regions, but the promise of atomic order, as exhibited in the momentary formation of ions, is there.

Somewhere between the hot interiors and the cooler envelopes of stars, the atomic order begins to prevail; the random motions of the electrons and nuclei are much more subdued, and the electromagnetic force now arranges these particles into atomic structures that may be pictured as consisting of a twofold order: the subatomic nuclear order of the neutrons and protons and the atomic order of the electrons revolving in their orbits around the nuclei. A complete atom thus represents a higher form of order than does the nucleus itself; the reason for this twofold order is that the electromagnetic force imposes an ordered superstructure of electrons on top of the nucleus. Thus, as additional forces come into play between basic particles of matter, more complex structures with higher forms of order and symmetry arise.

The electromagnetic force has so many facets that it can and does give rise to an incredibly complex variety of structures. The properties of all molecules, organic and inorganic; aggregates of molecules; crystalline substances, such as sugars, salts, ice, and snowflakes; glassy materials; and liquids can be accounted for by the electromagnetic force and by the many variations of this force, such as friction, that at first sight appear to be new forces and quite unrelated to electricity and magnetism. Thus, such properties of matter as hardness, stickiness, brittleness, density, sweetness, fragrance, and color, to name a few, are determined by electromagnetic forces. The nuclear force is so strong that it remains in control of the nucleus over a very large temperature range, which is quite fortunate, for otherwise the nuclei of atoms would be unstable and transmutations of chemical elements would occur quite readily with relatively small changes of temperature. The electromagnetic force, however, retains control of a structure only over a relatively narrow range of temperatures; as a general rule, the more complex the atomic or molecular structure, the narrower is the temperature range within which it is stable. This is already indicated by the difference between the stability of the simple atoms hydrogen and helium and the complex, heavy atoms like iron and calcium. Whereas helium and hydrogen remain un-ionized even at fairly high temperatures—up to about 10,000°K—the heavy complex atoms begin to lose their electrons at a few thousand degrees. That is why the atmospheres of the relatively cool orange and yellow stars contain the ionized atoms of the metals but the un-ionized atoms of hydrogen and helium. The atmospheres of the hot white and blue-white stars, however, contain the ionized atoms of hydrogen and helium as well as the manifold ionized atoms of the heavy elements.

Clearly the complexity of the order that prevails on the atomic level is determined not only by the electromagnetic force but also by the ambient temperature or the violence of the random motions of atoms and their constituent particles. Higher and higher forms of atomic order set in as the random thermal kinetic energy of nuclei and electrons diminishes and the electromagnetic force binds these particles together.

The Quantization of Atomic Order

Because the various levels of atomic order that exist in nature are determined by the electromagnetic force and the temperature, as the temperature of a mixture of electrons and nuclei drops, the complexity of the atomic order increases; the negatively charged electrons are bound in definite dynamical configurations by the electrical attraction of the positively charged nuclei. However, if the electromagnetic force and the temperature were the only physical entities involved in determining the electronic configurations, and therefore the degree of electronic order inside atoms, there would be no hierarchy of atomic structures nor a sequence of distinct chemical elements such as exists. The most remarkable property about the chemical elements, which is a consequence of the electronic order inside atoms, is that if they are arranged in a sequence of increasing mass—that is, by atomic weight, which is essentially the sum of the number of neutrons and protons in the nucleus of the atom—starting with hydrogen and ending with the heavy radioactive elements like uranium, the chemical properties and characteristics of the elements repeat themselves periodically along this sequence. Thus, the gases helium (atomic weight 4), neon (atomic weight 20), argon (atomic weight 40), krypton (atomic weight 84), xenon (atomic weight 132), and radon (atomic weight 222) are all chemically inert—the family of the so-called noble gases. The metals lithium (atomic weight 7), sodium (atomic weight 23), potassium (atomic weight 39), rubidium (atomic weight 85), cesium (atomic weight 133), and francium (atomic weight 223) constitute a family of chemically similar elements known as the alkali earths; they are extremely active and combine readily with many elements. These are but two of the dozen or so families that exist.

That all the chemical elements can be grouped into families, each of which is characterized by a distinct set of chemical properties, indicates that the electrons, which determine the chemical properties of atoms, are arranged around the nuclei in definite dynamical patterns, orbits, or configurations that repeat themselves periodically along the sequence. Indeed, if such definite dynamical electron patterns did not exist inside atoms, distinct, stable chemical elements could not exist, and hence, there would be no chemistry and no life. Thus, the presence of life in the universe can be traced back to the existence of immutable dynamical electronic patterns inside atoms. If, then, one is to understand how the organic molecules that constitute all living creatures were formed from the basic chemical elements in the universe, one must understand why definite electronic patterns exist inside atoms.

At first sight, the existence of such patterns may not seem to be difficult to explain in terms of the basic Newtonian concepts of force and motion, but in fact it is contrary to—or, more accurately, cannot be explained by—the classical (Newtonian) laws of physics; owing to this, the stability and unvarying chemical properties of atoms were extremely puzzling to chemists and physicists, but all of this was changed and the problem of atomic structure was solved with the formulation of the quantum theory by Planck in 1900 and the application of quantum concepts to the structure of atoms by Bohr in 1913. To see why the concept that electrons are arranged in precise, immutable patterns around the nuclei of atoms conflicts with the Newtonian laws of motion, one must consider the dynamical pattern of the planets around the sun and compare it with the electronic pattern in an atom like carbon. The planets move around the sun in well-defined, stable orbits that can be deduced from Newton's law of gravity combined with his laws of motion. The orbit of each planet as deduced from Newton's laws is an ellipse, which is characterized by three physical quantities: (1) the size of the orbit, or its major axis, which is the equivalent of the largest diameter of the orbit, the sum of the perihelion and aphelion distances of the planet, or twice the mean distance of the planet from the sun; (2) the shape, or eccentricity of the orbit, measured by the area enclosed by the orbit (the rounder the orbit of a given size is, the greater the area; the narrower or more elliptical the orbit of the same size, the smaller the area); (3) the orientation of the plane of the orbit to the sun's equator or to some other plane such as the plane of the earth's equator. One can deduce from Newtonian theory that the size of the planet's orbit is a measure of the total energy, kinetic and potential, of the planet, and the shape is a measure of the planet's rotational motion or angular momentum. A fourth dynamical quantity can be associated with the motion of a planet if one takes into account its rotation or spin and considers the orientation of the planet's axis of rotation with respect to some chosen direction. Thus, the tilt of this axis with respect to the line that is perpendicular to the plane of the planet's orbit—23.5° in the case of the earth—can be chosen for the purpose. One may thus picture a set of four numbers associated with each planet that specify the dynamical patterns of the planets, but there is nothing in Newtonian theory that says that this set of four numbers for each planet cannot be altered to any desired degree.

If a computer were given all the initial data about the states or conditions of the planets when they were first formed, then, according to Newtonian theory, the computer could, in principle, deduce the orbits of the planets and obtain the four dynamical numbers now associated with each planet. But in Newtonian theory it is possible to have a solar system just like earth's, with the same sun and the same planets, but with each planet having a somewhat different set of four numbers. Newtonian theory places no restrictions on the sizes, shapes, and orientations of the planetary orbits other than adherence to the laws of motion; it allows for the existence, without restriction, of planetary orbits of any size, shape, or orientation, so that there is no fixed pattern for a planetary system according to Newtonian laws. The pattern that emerged after the solar system was born depended only on the state of motion of the planets at that initial moment; had these been slightly different, the present system would also be slightly different. This seems to be borne out by observational data that show comets and other bodies in the solar system moving in a great variety of orbits.

Compare this now with the fixed electronic pattern in the structure of atoms, using the carbon atom as an example. The tightly packed nucleus of this atom contains 6 neutrons and 6 protons; its atomic weight is thus 12, and it has just 6 units of positive electric charge, one for each proton. Circling around the nucleus of the electrically neutral carbon atom are just 6 negatively charged particles, the electrons, forming a system that is dynamically similar to the solar system. There is, however, one vast and extremely important difference between the two: the dynamical pattern of the electrons in the carbon atom is a definite one, always the same regardless of how the 6 electrons were moving before they were captured by the carbon nucleus to form the complete carbon atom. This may be expressed somewhat differently as follows: Picture each one of a large number of carbon nuclei capturing 6 electrons from a vast number of randomly moving electrons, and then examine the arrangement of the electrons inside each carbon atom that is thus formed. One finds that the arrangement of these electrons in each atom is identical; the dynamical electronic pattern is the same inside every carbon atom, no matter how it was formed from a carbon nucleus and 6 electrons.

Just as a different set of four dynamical numbers can be used to describe the motion of each planet in its orbit around the sun, so it is with each electron inside an atom. However, whereas no restriction is placed by Newtonian theory on the values of the numbers in a set assigned to any planet, nature places very definite and remarkable restrictions on the four numbers assigned to each electron inside an atom. These restrictions are implicit in the quantum theory, which must be used instead of the Newtonian laws of motion to describe the motions of electrons inside atoms. To each electron inside the carbon atom can be assigned a set of four numbers that represent the size of its orbit (its total energy), the shape of its orbit or the area enclosed by the orbit (its angular momentum), which is uniquely related to the shape, the tilt of its orbit, and the tilt of the axis of spin of the electron with respect to the plane of its orbit. The profound difference between the set of four numbers that the quantum theory assigns to an electron and that Newtonian theory assigns to a planet is that the set of four electronic numbers, called quantum numbers by physicists, is a set of integers that are related to each other in a definite way, whereas the four planetary numbers are not restricted to related integers but can take on all values. Since the quantum numbers in any set are all integers, which constitute a discrete set, there can only be discrete sequences of such sets of numbers for any electron. This means that the orbits of electrons inside an atom form a discrete set. It is because of this discreteness that one speaks of the quantization of the motion of electrons inside an atom, and this quantization, together with another remarkable principle to be presented below, gives rise to definite, unchanging electronic patterns inside atoms and definite chemical properties that are repeated cyclically among different chemical elements.

Quantum Theory and Atomic Structure

To see the full significance of the quantum numbers and how they lead to discrete dynamical patterns inside atoms, one must show why the quantum theory was necessary and give a brief description of how it was introduced. During the last decade of the nineteenth century, physicists found that the observed quality of radiation (the measured amount of energy, or the measured intensity of the radiation, concentrated in each color) emitted per second from a small hole in a hot furnace did not agree with the theoretical results deduced from the classical wave theory of radiation and classical thermodynamics. These theories predict that most of the emitted energy, regardless of the temperature of the furnace, should come out in the form of ultraviolet radiation. But this is not the case at all: very little energy is emitted in the ultraviolet part of the spectrum unless the temperature of the furnace is extremely high. One finds, in fact, that the emitted energy is distributed among the various colors of the spectrum in such a way that the maximum intensity of the radiation is concentrated in one particular color that depends on the temperature. As the temperature varies from 1,000° to about 10,000°, the color of maximum intensity shifts from red to blue. That is why the interior of a cool furnace looks reddish and that of a very hot furnace looks white hot. Planck showed that as long as one assumes that the hot furnace emits its radiant energy continuously—that is, in the form of a continuous electromagnetic wave—one always obtains a theoretical result that does not agree with the observations. He then went on to show that perfect agreement can be obtained with the observations if one assumes (now no longer an assumption but a fact) that the furnace emits its energy discontinuously in the form of tiny energy pellets, or "quanta" —hence the name quantum theory.

To make this quantum theory fit the facts, Planck proposed a remarkable relationship between the amount of energy concentrated in a quantum of radiation and the frequency of that radiation (the number of electromagnetic vibrations per second, which determines the color of the radiation; the redder the light the lower the frequency; the bluer the light the higher the frequency). He showed that the higher the frequency—that is, the bluer the radiation— the more energy is packed into a quantum of energy, so that each blue quantum contains about twice as much energy as each red quantum. This is expressed as follows: Let E be the energy of a given quantum of radiation and f its frequency, then $E = hf$, where h is the famous Planck universal constant of action. This remarkably simple, almost magical, relationship between energy and frequency, known as the Planck equation, is the basis of the quantum theory. Planck's discovery of the constant of action h must rank in importance with Newton's discovery of the universal gravitational constant and with the discovery of the constancy of the speed of light, which led to the theory of relativity. Even though the numerical value of h is exceedingly small—it equals 6.7 divided by 1,000 trillion trillion erg seconds—it plays a dominant role in the structure of all matter and accounts for the difference between the existing complex, variegated universe of matter and a universe devoid of variety and chemistry. If the constant h were suddenly to become zero, the quantum theory would pass over to classical theory with tragic results: the electrons inside all atoms would collapse onto their nuclei, molecules would disappear, and life would be impossible.

To understand why the constant of action h plays so important a role in the structure of matter, one must note two things: (1) h is the basic unit of action in nature; and (2) the action of a dynamical system can only change by integral multiples of h. It is because of this discreteness of action that all atoms and molecules have definite, stable structures. The easiest way to demonstrate this is to consider the action associated with a single particle—for example, an electron—moving in a circular orbit around a fixed attracting central body like a nucleus. If the speed of the particle in its orbit is v and its mass is m, the total action that the particle has in its orbit is the product mvq of its mass m, its speed v, and the length q of the circumference of its orbit. This definition of action shows that only a discrete set of sizes of circular orbits can exist: the smallest orbit is that for which mvq just equals h (one unit of action); there can be no orbit smaller than this, because h is the smallest amount of action that can exist. Thus, electrons cannot collapse onto their nuclei. Orbits larger than this smallest one can exist, but only such sizes are permitted for which mvq has the values $2h$, $3h$, $4h$, and so on; in other words, only a discrete set of sizes is allowed.

In 1913 Bohr applied these ideas to the hydrogen atom and thus introduced modern atomic theory, which is the basis of modern chemistry. He showed that if e is the electric charge on the electron and m is its mass, then the only circular orbits for the electron that are permitted are those whose radii are $n^2h^2/4\pi^2e^2m$, where n is any one of the positive integers 1, 2, 3, ... The smallest orbit is thus one whose radius is $h^2/4\pi^2e^2m$. The integer n is called the principal quantum number of the electron; it is a measure of the electron's energy when the electron is in the nth orbit.

The principal quantum number specifies only the size of the electron's orbit (the energy), but tells nothing about the various possible orbital shapes (angular momentum). If electronic orbits inside atoms were always circles, only the principal quantum number would be needed to specify the orbits, which would differ only in size; but noting that electronic orbits, like planetary orbits, are ellipses, with a great range of shapes, from a very flat ellipse to a circle, which is a special case of an ellipse, one must consider all the possible orbital shapes associated with an orbit of given size. Newtonian physics places no restrictions on such shapes, so that in classical physics there can be infinitely many orbits of the same size (same energy) but of different shapes (different angular momentum). This is not so according to the quantum theory, because a change of shape means a change in angular momentum and hence a change in action; since action can only change in steps of h, only those orbital shapes are permitted for which the action of the electron associated with its angular momentum is an integral multiple of h. Using the quantum theory, one can show that the number of permissible orbits for an electron, all of which have the same size but different shapes, is equal to the principal quantum number that determines the size of these orbits. This simply means that there are exactly n orbits, each with a different shape, associated with the principal quantum number n. Thus, there is only 1 orbit that has the principal quantum number $n = 1$; there are 2 differently shaped orbits (ellipses), each with the principal quantum number $n = 2$; there are 3 orbits, each of different shape, with the principal quantum number $n = 3$;

and so on. Each of the n shapes associated with the principal quantum number n is specified by one of the integers from 0 to $n - 1$. The flattest shape is assigned the number 0 and the roundest shape is assigned the number $n - 1$. Thus, the 3 orbital shapes associated with the principal quantum number 3 are designated 0, 1, 2.

But this is not the whole of the quantization picture, for it can be shown from the quantum theory that the possible orientations of each of the variously shaped orbits with respect to some fixed direction in space—for example, with respect to the direction of the lines of force of a magnetic field in the space between the north and south poles of a magnet—are not arbitrary but are restricted to a discrete set of numbers that are determined by the quantum number assigned to the shape of the orbit. In fact, the possible number of discrete orientations of an orbit of given shape is 1 plus twice the quantum number l that designates that shape, and each of these orientations is specified by one of the integers (zero included) between $-l$ and $+l$, which are called orbital, or azimuthal, quantum numbers. To be specific, consider an electron in an orbit whose principal quantum number n is 3, where the size of the orbit is $9h^2/4\pi^2e^2m$, and whose shape is given by the azimuthal quantum number $l = 2$; according to the quantum theory, this orbit can have any one of five orientations $(1 + 2 \times 2)$ specified by the five integers $-2, -1, 0, 1, 2$ (called magnetic quantum numbers m). Finally, the spin axis of the electron can have one of two perpendicular, but opposite, orientations with respect to the plane of the electron's orbit around the nucleus, which may be represented by the spin quantum number s and which may be set equal to $+1$ for clockwise spin and -1 for counterclockwise spin. One thus sees that an electron moving in some orbit around the nucleus of an atom does not obey the Newtonian laws of motion, which place no restrictions on the size, shape, or orientation of its orbit or spin, but obeys instead the quantum laws of motion, which restrict the orbit to one of a discrete set of sizes, shapes, and orientations specified by definite sets of integer quadruplets. It is this discreteness, as defined by the four integer quantum numbers in each set, that leads to definite electronic orbits and therefore to definite stable atomic structures and definite molecular structures, no matter under what circumstances the atoms and molecules are formed. From the description of the various quantum states of an electron inside an atom, one sees that, even though the electron is restricted to sets of discrete states, there is such a variety of these that many different stable atomic structures are possible; therefore, different chemical elements are also possible.

Since the description here of the way the various quantum numbers in a set are assigned to an electronic orbit has been somewhat general, the procedure will be illustrated by a few specific examples. Since the smallest possible electronic orbit in an atom is given by the principal quantum number $n = 1$, only one orbital shape (azimuthal quantum number $l = 0$) is possible. Since $l = 0$, this orbit has only one possible orientation given by the magnetic quantum number $m = 0$. Thus, the ground state of an atom or the lowest dynamical electronic configuration with the smallest size orbits consists of only one orbit with quantum numbers given by $n = 1$, $l = 0$, $m = 0$ if the spin of the electron is not taken into account. Taking note, however, that an electron can spin either clockwise or counterclockwise in such an orbit, one can assign either the spin quantum number $s = +1$ or $s = -1$ to the electron. There are thus two, and only two, distinct electronic vibrational or orbital configurations associated with the ground state (principal quantum number $n = 1$), which is called the K state, or K electronic shell, of an atom. One orbital configuration is given by the four quantum numbers $n = 1$, $l = 0$, $m = 0$, $s = +1$ and the other by the quantum numbers $n = 1$, $l = 0$, $m = 0$, $s = -1$. Notice that three of the quantum numbers in each set are alike but one is different.

The next higher set of orbits, which lie above the K shell and constitute the L shell of the atom, all have the same size, given by the principal quantum number $n = 2$, and there are two different orbital shapes associated with this size because $n = 2$. The narrower of these shapes is specified by the azimuthal quantum number $l = 0$, and the rounder shape is given by $l = 1$; the $l = 0$ shaped orbit has just one orientation, given by the magnetic quantum number $m = 0$, but the $l = 1$ shaped orbit has three orientations, $m = -1, 0, +1$. There are thus four distinct possible orbital configurations of the electron in the L atomic shell if one does not take into account the spin of the electron, but since there are two possible spin orientations given by the spin quantum number, $s = +1$ and $s = -1$, the total number of distinct possible orbital configurations is eight. Each of these is designated by its own set of four quantum numbers; no two sets of quantum numbers are identical. The arrangement of the electrons in the K and L shells of the carbon atoms is shown in the diagram.

In the higher shells, with larger principal quantum numbers, the variety and number of electronic orbital configurations increase; thus, for $n = 3$ there are 18; for $n = 4$ there are 32; and so on. In general, for any principal quantum number n there are just $2n^2$ distinct electronic orbital configurations or vibrational states. It is this consequence of the quantum theory and the quantization of action that leads to the different chemical elements, to molecular structures, and to chemical interactions in general. But one more physical principle must be introduced before one can understand why, even with the discrete electronic configurations demanded by the quantum theory, different chemical elements that can be arranged in chemically similar families exist. The existence of discrete atomic shells and discrete electronic configurations in each shell does not tell why all the electrons in any given atom do not drop into, and stay in, the lowest shell, the K shell. If one could examine the carbon atom with a supermicroscope,

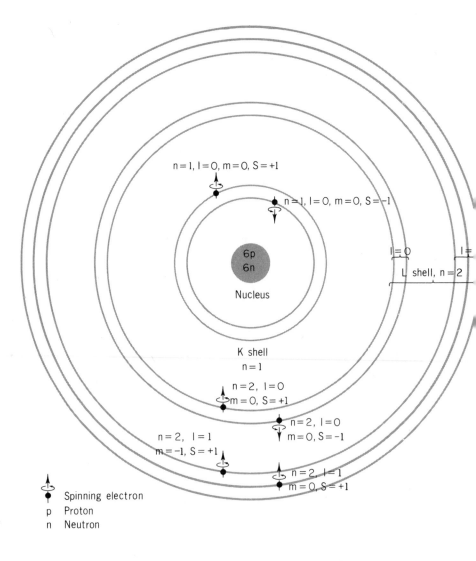

The arrangement of the six electrons in the carbon atom. Two electrons are in the K shell with opposite spins; they have the same quantum numbers $n = 1$, $l = 0$, $m = 0$. The four other electrons are in the L shell; two of these are in orbits $l = 0$, $m = 0$, but with opposite spins, and the other two are in orbits $l = 1$, $m = -1$, and $l = 1$, $m = 0$, but with parallel spins as shown. This arrangement of the four outer electrons gives the carbon atom its four fold valency and hence its great chemical versatility.

one would find that only two of its six electrons circling the nucleus are in the K shell; the other four are in the L shell, and they stay there. This is quite contrary to what one would expect from Newtonian physics and even from quantization alone.

To see what the problem is from the point of view of classical theory, consider a crater in the earth with six identical spherical boulders along the rim of the crater. If the boulders are pushed over the rim of the crater, each boulder will roll down as far as it can and come to rest as close to the bottom of the crater as possible; but the six boulders when they have come to rest will display no particular, predictable pattern, and this is what one expects from Newtonian theory. If the six boulders are brought back to the rim of the crater and allowed to roll down the slope of the crater again, they will arrange themselves close to the bottom of the crater just as before, but the boulder pattern will be somewhat different from what it was the first time. In fact, there would be a slightly different boulder pattern at the bottom of the crater, in agreement with Newtonian theory, each time the experiment was repeated, because the quantum theory, although operative here too, has only a very minute effect on the behavior of the massive boulders; the action of a boulder is so large that many millions of millions of multiples of h are involved in its motion, so that a change of a single unit of action h in its total action is completely unobservable.

To draw an analogy with the behavior of electrons inside atoms, one can introduce an artificial quantization of the boulder pattern by picturing concentric circular and elliptical channels dug around the walls of the crater. Picture only two such deep concentric channels with equal diameters near the very bottom of the crater; eight equally sized, differently shaped channels, with larger diameters than the first two, higher up on the sides of the crater; and so on. If only one boulder is now allowed to roll down the slope of the crater, it will end up in one of the two lowest channels, and this would be true of all six boulders if one did not impose the restriction that a single channel can accommodate one and only one boulder. With this restriction the six boulders would always arrange themselves in the same pattern inside the crater because only two boulders would end up in the two smallest channels and the remaining four would then, perforce, occupy four of the second set of channels. If one carries this picture over to six electrons being captured by a carbon nucleus to form a neutral carbon atom and imposes the same kind of restriction on these electrons, one can account for the dynamical configurations of the six electrons in such an atom: two move around the nucleus in the K shell and four in the L shell. This restriction is imposed by the Pauli exclusion principle, discovered in 1927 by the great Austrian-born Swiss physicist Wolfgang Pauli (1900–1958). This principle states that only one electron can move in a given orbit inside an atom; when one electron is in any one of the orbits, all other electrons are excluded from moving in that orbit. This principle may be stated somewhat differently by saying that no two electrons inside the same atom can have the same set of four quantum numbers. It is clear from this remarkable principle and the existence of discrete sets of quantized orbits that the electrons inside any atom of a given type must fall into the same pattern no matter how they were captured by the nucleus of that atom, and this pattern—particularly the pattern of the electron in the outermost shell—determines the chemical properties of that particular kind of atom.

A good deal of space has been devoted to the discussion of quantization and how quantization leads to the dynamical electronic patterns inside atoms, because without such definite, fixed, quantized patterns, atoms could not combine to form definite complex organic molecules, and life itself would be impossible. Life can exist only because the organic molecular structures that are the basis of all life are stable and can only change to new configurations by discrete transitions involving fairly large, discrete quantities of energy instead of by continuous absorption of minute amounts of energy, as would be the case if there were no finite quantum of action. Consider, for example, a gene, one of a number of large organic aggregates of atoms in the nucleus of a living cell that are essential not only for the continuance of life but also for the propagation and perpetuation, from generation to generation, of the definite physical and chemical characteristics of a living species that distinguish the species from all others. That a particular characteristic has remained essentially unaltered for thousands of generations points to the amazing stability of the gene through very many successive cell divisions even though it is subject to the constant buffeting of other molecules in a cell, which is known as thermal motion. Even more striking is an example given by the Austrian physicist Erwin Schrödinger (1887–1961), the father of wave mechanics; it illustrates the remarkable stability of a single gene responsible for the continuous perpetuation of a single physical feature in a given human family through many generations. The portraits of various members of the Habsburg dynasty from the sixteenth century down through the nineteenth century show an unusually large protuberance of the lower lip that is known to geneticists as the Habsburger lip. It is clear from the persistence and continual reappearance of this facial characteristic that the highly ordered structure of the single gene that is responsible for it is reproduced faithfully, generation after generation, in spite of the disordering tendency of the thermal motion to which the gene has been subjected for centuries. Since the genes in the human body are maintained at a constant temperature of 98.6°F and are therefore constantly buffeted by thermal motions, it is impossible to explain their unchanging structures on the basis of classical physics.

A nineteenth-century physicist, confronted with this problem, would have explained it by stating (correctly) that genes must be organic molecules, since molecules are stable. But this would really have been no explanation without an explanation of the stability of molecules; but this explanation came only after the introduction of the quantum theory, which is the key to the mystery of all molecular structures. Since the stability of all organic molecules and the very mechanism of heredity are based on the quantum theory, one must first see how the quantum theory reveals the structure of molecules before one can tackle the problem of the origin of life.

Quantum Theory and Molecular Structure

The following question immediately confronts one when considering a molecule such as ordinary table salt, sodium chloride (NaCl), or the hydrogen molecule (H_2): How is it possible for two neutral atoms, such as an atom of sodium and an atom of chlorine or two atoms of hydrogen, to stick together to form a molecule? Classical physics cannot answer this question, because neutral atoms appear to be incapable of exerting any force but the gravitational force, which is so minute for particles of such small mass that it can be completely disregarded. Indeed, if two neutral atoms are brought close to each other, one would expect them to repel each other because the negatively charged outer electrons of one of the atoms would repel those of the other and the two positively charged nuclei would also repel each other. Instead, the two atoms, if properly chosen, stick to each other to form a molecule.

Before the discovery of the quantum theory and the quantization of electronic configurations inside atoms, the question of the nature of chemical bonds between two atoms could not be answered, but it is now known that there are two kinds of chemical bonds: the ionic, or heteropolar, bond, exemplified by a molecule of sodium chloride, with two different atoms; and the homopolar, or covalent, bond, exemplified by the hydrogen molecule, with two identical atoms. The ionic bond can be understood if one notes that a molecule of salt in solution does not break up into two neutral atoms (Na and Cl) but into two electrically charged particles: a positively charged sodium ion (Na^+) and a negatively charged chlorine ion (Cl^-). The formation of a neutral molecule of salt can now be pictured as occurring as follows: The loosely bound lone electron in one of the $n = 3$ orbital states of the sodium atom leaves this atom to join the seven electrons moving in seven of the $n = 3$ orbital states around the chlorine nucleus. The sodium atom thus becomes a positively charged ion and the chlorine atom becomes a negatively charged ion, and the two oppositely charged ions stick together owing to their mutual electrostatic attraction. There is no mystery here, but even so, classical physics cannot explain why the sodium atom loses its electron and the chlorine atom swallows up the lost electron. It can only be understood on the basis of the four quantum numbers and the discrete electronic configurations that stem from these quantum numbers. The quantum theory shows that an atom with less than half of the allowed number of electrons permitted in its outermost shell tends to lose one or more of these to an atom that has more than half of the allowed number of electrons permitted in its outer shell. It is this property that in a general way accounts for the chemical affinities of atoms. Carbon is of special interest because the number of electrons in its outer shell is just four, which is half the number permitted in that shell. Thus, carbon can absorb up to four electrons and also lose up to four electrons, so that it has an enormous versatility in its affinity for different kinds of atoms; this versatility is the basis of organic chemistry and of life itself.

A simple example of a heteropolar bond is given here to show how the number of electrons in the outermost shell of an atom determines its chemical properties and its affinity for other atoms. Consider hydrogen and oxygen; hydrogen has only one electron circling its nucleus in the K shell, but oxygen has eight electrons circling its nucleus, with two in the

K shell and six in the L shell, which has room for just eight electrons. Thus, oxygen will tend to collect two electrons from other atoms whenever it can to fill its outer shell, and so, if two hydrogen atoms are brought close enough to a single oxygen atom, the two electrons of the hydrogen atoms will be strongly attracted to the oxygen atom to complete the oxygen atom's L shell and the two hydrogen atoms will thus be bound to the oxygen atom to form a molecule of water (H_2O) by sharing their two electrons with the oxygen atom. Although this is a rough description of how different kinds of atoms combine to form molecules, it is essentially correct and provides a good rule of thumb for determining the valence of an atom, the number of hydrogen atoms it combines with to form a molecule.

Consider now a homopolar bond. The normal states of such gases as hydrogen, nitrogen, and oxygen are not the atomic states H, N, and O, but the molecular states H_2, N_2, and O_2. To see why this is so, observe two hydrogen atoms brought close together. Whether they repel each other or attract each other and thus combine to form a molecule depends entirely on the two electrons that are spinning. If the two electrons are both spinning clockwise or both spinning counterclockwise, the two atoms repel each other and the H_2 molecule is not formed; but if one is spinning clockwise and the other counterclockwise, the atoms attract each other and are then bound together to form the H_2 molecule. The way the atoms attract each other is a purely quantum mechanical phenomenon; the two electrons, being indistinguishable, behave as though they were exchanging places at a certain frequency, and this exchange binds the two atoms together. It is as though the two protons were juggling the electrons between them and were held together by this juggling act; the two protons share the two indistinguishable electrons. It is important to note that because of quantization, in both the ionic bond and the covalent bond only definite stable molecular patterns can occur, and this insures molecular stability. A molecule cannot change its structure gradually but only in discrete steps that require discrete amounts of energy. Ordinary thermal motion does not have enough energy for this.

Organic Molecules

Without carbon there would be no organic chemistry and hence no life. The reason for this is that there are just four electrons in the L shell of the carbon atom, so that carbon has a valence of four; this means that it can combine with four hydrogen atoms to form the hydrocarbon molecule methane (CH_4), which is one of the simplest organic molecules. Although this molecule, simple as it is, already shows the great molecular diversity that one can expect among carbon compounds, the incredible variety and richness that actually exist are revealed only when other atoms such as oxygen, nitrogen, and phosphorus also come into the picture. Because carbon atoms can form bonds not only with these other atoms but with themselves as well, large organic molecular chains can be built up in which the interchange in the position of two different atoms or any change in the spatial arrangement of the atoms gives rise to a different organic compound. Chemists refer to two different compounds that consist of the same group of atoms arranged differently as isomers. Isomerism is not the exception among organic compounds but the rule; the larger the organic molecule, the larger is the number of possible isomers.

The way in which the atoms in organic compounds arrange themselves to form isomers can be presented graphically by picturing the carbon bonds by means of four dashes, each representing a single valence, extending out of the carbon atom as follows:

$$-\overset{\displaystyle |}{\underset{\displaystyle |}{C}}-, \text{ or } =C\!\!=\!\!, \text{ or } \overset{\displaystyle |}{\underset{\displaystyle |}{C}}\!\!=\!\!,$$

and so on. All or any of these four bonds can be pictured as being shared by some other atom, which itself can be pictured with its own bonds represented as dashes. Thus, the carbon monoxide molecule (CO) is pictured as $=$C$=$O, with the single oxygen atom sharing only two of the bonds. The presence of two unshared, or unsaturated, carbon bonds in this molecule makes carbon monoxide extremely volatile and poisonous because it readily unites with any free oxygen and with the hemoglobin in the blood. When carbon monoxide burns in air, it forms carbon dioxide (CO_2), which may be written graphically as O$=$C$=$O. Although such structural formulas are not to be taken too literally as indicating the true spatial arrangement of the atoms, they are useful in helping one understand some of the properties of organic compounds. Thus, from the fact that carbon monoxide has two unshared carbon bonds, whereas all four carbon bonds in carbon dioxide are shared, one deduces that carbon dioxide is much more stable and far less chemically active than carbon monoxide. Consider now the hydrocarbon methane (CH_4); its structural formula is as follows:

$$
\begin{array}{c}
\text{H} \\
| \\
\text{H—C—H} \\
| \\
\text{H}
\end{array}
$$

Note that each of the carbon bonds is attached to a single hydrogen atom. Since no carbon bond is unattached, such a hydrocarbon is referred to as saturated. The great versatility of carbon for forming complex molecules lies in the fact that carbon atoms can link themselves together with their own bonds to form long carbon chains to which many other kinds of atoms can attach themselves. Consider, for the moment, the group of hydrocarbons known as the n-alkanes (paraffin), whose general formula is C_nH_{2n+2}, where n can be any integer;

this group includes such compounds as ethane (C_2H_6), propane (C_3H_8), and butane (C_4H_{10}) with the following structural formulas:

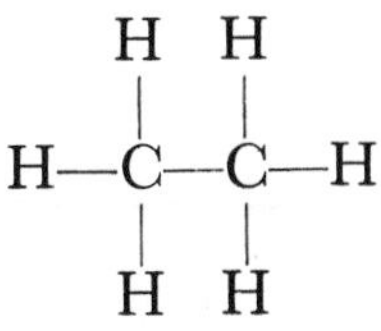

$$
\begin{array}{ccc}
\text{H} \quad \text{H} & \text{H} \quad \text{H} \quad \text{H} & \text{H} \quad \text{H} \quad \text{H} \quad \text{H} \\
| \quad\quad | & | \quad\quad | \quad\quad | & | \quad\quad | \quad\quad | \quad\quad | \\
\text{H—C—C—H} & \text{H—C—C—C—H} & \text{H—C—C—C—C—H} \\
| \quad\quad | & | \quad\quad | \quad\quad | & | \quad\quad | \quad\quad | \quad\quad | \\
\text{H} \quad \text{H} & \text{H} \quad \text{H} \quad \text{H} & \text{H} \quad \text{H} \quad \text{H} \quad \text{H} \\
\text{ethane} & \text{propane} & \text{butane}
\end{array}
$$

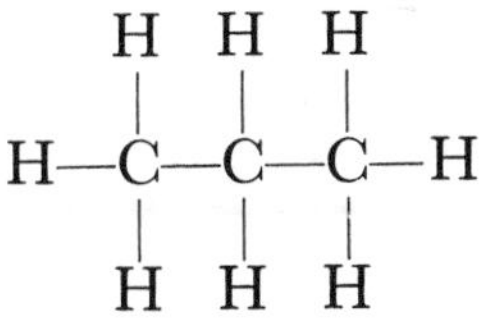

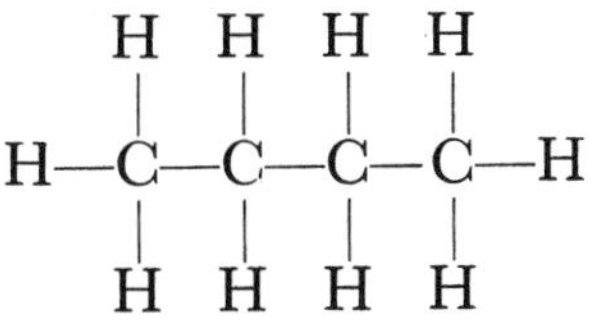

In these structural formulas the carbon atoms are arranged in a linear chain, but other arrangements that give isomers of these propane and butane hydrocarbons are possible. Thus, in addition to the usual butane above, there is isobutane, whose structural formula is as follows:

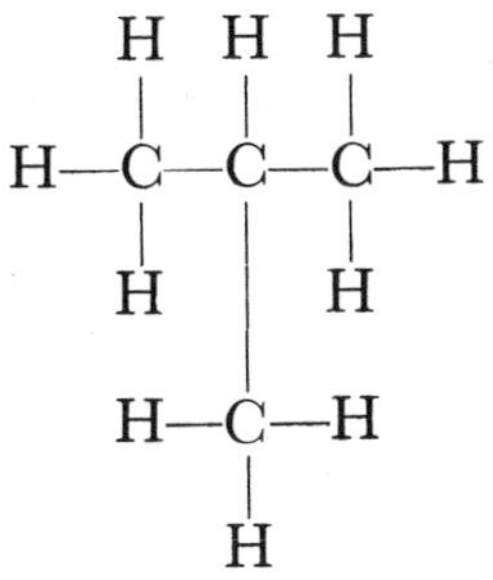

$$
\begin{array}{ccc}
\text{H} & \text{H} & \text{H} \\
| & | & | \\
\text{H—C} & \text{—C—} & \text{C—H} \\
| & | & | \\
\text{H} & & \text{H} \\
& | & \\
& \text{H—C—H} & \\
& | & \\
& \text{H} &
\end{array}
$$

An important group of hydrocarbons known as the benzenes, or the aromatic compounds, has a ring of six carbon atoms as its basic structure; in this ring the carbon atoms are arranged in a hexagon with alternate single and double bonds. In the benzene molecule itself a single hydrogen atom is attached to each one of the six remaining valences of the carbon atoms in the ring, so that all the carbon valences are saturated as shown here in the structural formula of benzene:

$$
\begin{array}{c}
\text{H} \quad \text{H} \\
| \qquad | \\
\text{C}\text{—}\text{C} \\
\diagup\!\!\diagup \qquad \diagdown\!\!\diagdown \\
\text{H—C} \qquad\qquad \text{C—H} \\
\diagdown \qquad\quad \diagup \\
\text{C}\!=\!\text{C} \\
| \qquad | \\
\text{H} \quad \text{H}
\end{array}
$$

benzene

The benzene rings are important because their great symmetry makes them extremely stable and able to act as units in forming more complex organic compounds such as the dyes; in such large molecules individual benzene rings attach themselves to each other.

An important group of organic compounds that contain oxygen atoms in addition to hydrogen and carbon is the alcohol group. Alcohol molecules are obtained from the hydrocarbons by replacing one of the hydrogen atoms by an oxygen atom. Thus, the two isomers of propyl alcohol are obtained by replacing one of the hydrogens in the propane molecule by an oxygen atom:

$$
\begin{array}{c}
\text{H} \quad \text{H} \quad \text{H} \\
| \quad\; | \quad\; | \\
\text{H—C—C—C—O—H} \\
| \quad\; | \quad\; | \\
\text{H} \quad \text{H} \quad \text{H}
\end{array}
\qquad\qquad
\begin{array}{c}
\text{H} \\
| \\
\text{H} \quad \text{O} \quad \text{H} \\
| \quad\; | \quad\; | \\
\text{H—C—C—C—H} \\
| \quad\; | \quad\; | \\
\text{H} \quad \text{H} \quad \text{H}
\end{array}
$$

propyl alcohol I $\qquad\qquad$ propyl alcohol II

Organic molecules have been discussed in some detail, with particular attention to isomers, because all of life is based on these molecules, a gene being a huge complex organic molecule that can only change discontinuously to form an isomer of itself by a rearrangement of its atoms. The examples of the simple isomers given here are important because they show that a single change in the arrangement of the atoms is sufficient to lead to an entirely different compound, but such changes do not occur spontaneously. Each isomer is stable and remains unaltered unless enough energy, in the form of a single quantum, is supplied to it to cause it to pass into another isomeric state. The importance of this for a gene is clear; genetic changes cannot occur continuously or spontaneously but only discontinuously, through the absorption of a quantum of energy that is large enough to cause a rearrangement of the atoms at some specific point in the gene. But such changes, or mutations, are extremely rare because even though vast numbers of isomers are possible in a single gene, each is a very stable configuration that is separated from any other isomeric configuration by a fairly large energy threshold. These isomeric energy thresholds in genes are high enough, compared to the average thermal energy of atoms in the human body, to make a transition from one genetic isomer to another a very rare event.

10 The Origin and Nature of Life

Quantum Theory and Stability of Life

The general description of the structure of atoms and molecules and the role that the quantum theory plays in determining and maintaining these structures shows why the molecular patterns around which all life is built are stable. The stability of life itself is essentially a consequence of the quantum theory, which, on the basis of first principles, accounts for all the kinds of atomic and molecular structures found in the universe. But one can infer more than that from the quantum theory: it also implies that, given enough time and the proper physical conditions, all such possible molecular structures will occur. The reason for this is found in the very discreteness and stability of the energy levels or energy configurations of molecular structures; once a particular aggregate of atoms has occurred and has settled down to its most stable state—a molecule—it becomes the starting point of a new set of stable states that can arise either by a rearrangement of the atoms in the aggregate or by the addition of another atom to form a more complex molecule.

To see what is involved here, consider some complex stable organic molecule all of whose atoms are arranged in a pattern of minimum energy. Suppose further that this molecule has no tendency to combine with other molecules or atoms in this lowest energy state. As long as this molecule is moving about in a region in which it is subjected to

236

perturbations so weak that they do not supply it with enough energy in a single quantum to change its atomic configuration, the molecule will remain as it is; it will not combine with other atoms or molecules to form more complex structures. But suppose that there exist higher energy levels of the given molecule such that when the molecule is in one of these higher energy states, it has a very great tendency to combine with some other atom or molecule. Suppose, further, that occasionally the molecule is jolted by a quantum of energy that is large enough to lift it into one of these discrete higher states of energy. The molecule then has a chance to combine with other atoms and molecules to form a more complex molecule. The reason for this is that after the original molecule has been excited to the higher state of energy by absorbing the proper quantum of energy, it does not immediately slip back to its lowest state of energy again but spends a certain finite time in the higher energy state, and this may just be long enough to allow the molecule to capture another atom or combine with another molecule. Thus, more and more complex molecules can be built up step by step, given the right conditions and a long enough period of time. This process can now be repeated, starting with the new, more complex molecules, so that molecules of still greater complexity can be built up step by step, with each new step differing from the previous one by a discrete, but not too large, amount of energy.

An analogy is introduced here to illustrate how, given enough time, discrete energy steps in atoms and molecules and discrete molecular patterns, as imposed by the quantum theory, permit the step by step, spontaneous synthesis of complex molecules from simpler ones through random molecular and atomic collisions, whereas this could not happen if classical physics prevailed. Imagine the pieces of a jigsaw puzzle moving about on a table randomly and colliding with each other rather gently, without meshing, as one would expect, even when they meet in just the right way. The reason for this is found to be that each piece on the flat surface is completely encased in a fairly tough transparent plastic skin that prevents a projection of one piece from meshing with an identically shaped indentation in another piece; the two pieces bounce off each other on colliding, because although the plastic casings stretch a bit, they do not break. As long as the random motions of these jigsaw-puzzle pieces are not violent enough to break their plastic skins when they collide, the pieces remain unattached and no buildup of pieces into more complex patterns occurs. This is to be compared to the random thermal motions of various organic molecules and atoms, which to be sure, are not surrounded by plastic membranes but which can exist only in discrete energy states; as long as the temperature is not very high—ordinary body temperatures—the collisions among them resulting from random thermal motions of such molecules do not disturb them enough to lift them from one discrete energy state to another.

Suppose now that every now and then, owing to some rare and unusually large perturbations, one or more of the jigsaw pieces on the tabletop acquire so much speed (they might have been hit successively three or four times in the same direction) that their plastic skins break when they collide with other pieces. The projections and notches of these pieces are now exposed, and linkages between such pieces can occur to form a more complex pattern. This complex pattern in turn can on certain occasions suffer additional penetrations of its remaining plastic skin and then link up with still other pieces to form an even more complex pattern. Given time enough, more and more pieces will link together to form large, well-defined patterns if just enough energy is available during a collision to break the plastic skin at appropriate points. In all of this it is important to note that nothing will happen unless just enough energy is available to penetrate the plastic skin. This is to be compared to the quantum of energy (analogous to breaking the plastic membrane) that is needed to lift a molecule from its ground state of energy to the next higher level, from which it can combine with another atom or molecule and thus arrange itself into a more complex and more highly ordered molecular structure. This quantum of energy can come from any energy source, such as a lighted match, the sun, or an electric current.

The Law of Disorder and the Persistence of Life

The description of the way complex organic molecules can be built up spontaneously from simpler molecules and atoms may leave one with the erroneous impression that the only thing that counts in this process is time and that given enough time, ever higher forms of order evolve spontaneously out of the original disorder; but this is not so, for in time all isolated systems of atoms and molecules reach a state of equilibrium in which no new events occur, so that synthesis of higher molecular forms stops. The word *isolated* here is crucial, for the only way a system of atoms and molecules avoids reaching a state of complete equilibrium—which for living systems means death—is by not being isolated, so that it can absorb energy and matter from some outside source and use this infusion to synthesize higher forms. One may illustrate this point by considering a system consisting of just two kinds of atoms, A and B, which can combine to form the molecule AB. But one must here take into account something not mentioned in the description of spontaneous systhesis: the spontaneous dissolution of the molecule AB into the atoms A and B. This happens because some of the collisions that the molecules suffer are sufficiently energetic to disrupt the molecule. In other words, two processes will be going on in molecular systems simultaneously: synthesis and dissolution. When the rate at which A and B combine to form the molecule AB is exactly equal to the rate at which this molecule dissociates into the two atoms, A and B, equilibrium prevails, and the further buildup of the molecule AB ceases. At this point in the spontaneous passage from complete disorder (a homogeneous mixture of the individual atoms A and B) to a state of partial order (a mixture of molecules AB and individual atoms A and B), the abundances of the atoms A and B and the molecule AB are fixed; to change these abundances something—either energy or matter—must be supplied to the system.

To see exactly what the mechanism of synthesis and dissolution is, suppose that one starts with a mixture of the atoms A and B only that is at a uniform temperature such that the average kinetic energy of these atoms is below, but not too far below, the energy threshold required for the formation of the molecule AB. In spite of this, some molecules AB are formed spontaneously because there are some individual atoms A and B in the mixture moving fast enough to combine when they collide; in the random atomic and molecular motions of a mixture there is always a definite distribution of velocities about an average value, which depends on the temperature of the mixture. In time, as more and more of the molecules AB are formed, the chance for the further buildup of these molecules is reduced, even though the temperature is kept the same, because the total number of the individual uncombined atoms A and B in the mixture is reduced; hence, the number of those moving fast enough to combine when they collide is also reduced, so that the rate of production of the molecule AB decreases. At the same time the rate of dissociation of the molecules AB back into the individual atoms A and B increases because as the AB population in the mixture increases, the rate of destructive collisions between them also increases. Thus, in time, if the temperature remains constant, a state of chemical equilibrium is reached and nothing more happens; the abundance of each of the species A, B, and AB remains constant. This can be changed by increasing the temperature or by introducing more of the atoms A and B, but unless that is done, stagnation—that is, complete homogeneity or a permanent state of equilibrium—occurs.

All this is a consequence of a very general natural

law—the famous second law of thermodynamics, which states that an isolated system of particles and energy will, in time, reach a state of maximum disorder; the general tendency in nature is toward homogeneity, equilibrium, and disorder and not toward differentiation, nonequilibrium, and higher forms of order. Now it may appear that this is contradicted by the very example of molecular formation A + B $\rightleftharpoons$ AB because it starts out with a mixture of individual, randomly moving atoms A and B, a state of complete disorder, and in time reaches a state of equilibrium in which, to be sure, the motion is again random but in which some molecules AB in addition to the atoms A and B are present. Is not this then a higher state of order than the original state, because molecules represent a higher form of order than do the individual atoms of which they are composed? The answer to this question is yes if one considers only the atoms and molecules in the mixture and leaves out of account the energy that was absorbed or released when the molecules AB were formed from the individual atoms. Energy also contributes to different states of order, and it can be shown that whether energy is released (exothermic chemical reaction) or absorbed (endothermic chemical reaction) in the formation of the molecule AB, the disorder produced by the absorbed or emitted energy is always greater than the order produced by the formation of the molecules. Indeed, whether a particular chemical reaction is possible or not depends on whether the total disorder— including the disorder of the energy—that remains after the chemical reaction is greater or less than that existing before the chemical reaction. Only if the total disorder increases will the chemical reaction occur.

The tendency of a completely isolated system to pass from order to disorder is measured by a physical quantity called the entropy of the system; the larger the entropy of a system, the greater is its state of disorder. One may, therefore, restate the principle that an isolated system tends to become more disordered as follows: The entropy of an isolated system can never decrease; the entropy of such a system tends toward a maximum and when the state of maximum entropy has been achieved the system is in equilibrium. Since the entropy of a system can be expressed in terms of the temperature and volume of the system, the entropy of the system, and hence its state of disorder, can be determined. In an isolated system that is changing, the entropy can remain constant only if the changes occur at an infinitesimal rate; since this is impossible in practice, the entropy and disorder must increase.

It may be useful here to cite a few examples of the spontaneous transition of systems from order to disorder or from differentiation to complete homogeneity. Consider two different gases in the same container but separated from each other by a partition. This is a differentiated state because the two sets of molecules are spatially arranged next to each other. If the partition separating the two gases is removed, the two gases quickly intermix until complete homogeneity prevails; no matter how long one waits, one will never find an ordered state occurring again in which the two sets of molecules are again separated and occupy different halves of the container.

As a second example, consider a hot piece of metal in contact with a piece of ice. In time the metal cools off and the ice melts, indicating that heat has flown from the hot substance to the cold one. Here the energy starts out in a differentiated state and then moves in such a way that in time a homogeneous distribution of the heat occurs. The heat is never found flowing from the ice to the piece of metal nor does a piece of metal immersed in water become hotter and hotter while the water becomes colder, ultimately freezing and becoming ice.

To understand what all this has to do with the existence and persistence of life, note first that the essential characteristic of life is its ability to create order out of disorder; life thus appears to deny the increase-of-entropy principle. Living objects, from the simplest one-celled creatures upward, are not in states of equilibrium but in highly ordered, differentiated states and maintain themselves in these highly differentiated states against the demands of the second law of thermodynamics. Is, then, the existence of living matter an example of the failure of this law; and if not, how does the living organism evade the decay to equilibrium? There is really no mystery here, and one need not call upon nonphysical or supernatural forces to account for the persistence of life in spite of the second law if one keeps in mind that the law of entropy applies only to isolated systems, and the condition of being isolated is far from being true of living organisms. In fact, if any living matter were completely isolated from its surroundings so that it could absorb neither energy nor matter from its environment, it would quickly decay into an inert state of equilibrium and become as disordered as inanimate matter. The living organism, which represents the highest state of order in nature, avoids this precisely because it is not isolated and is thus able to exchange the disorder, which stems from its living processes, for the order that it absorbs from its environment; this process is called metabolism and assimilation. The living organism assimilates atoms and molecules into its own being by eating, drinking, and breathing; but more important, it does so in such a way as to decrease its own total entropy and to maintain its high state of order. Everything that a living organism does creates disorder, or positive entropy, but by metabolizing and assimilating food, which is matter in a very high form of order, the living organism is able to get rid of its positive entropy and maintain and even increase its own state of order. The net result, in spite of the order achieved by the living organism, is an increase in the total entropy in the universe, so that life at any one place in the universe maintains itself by feeding on the order, or negative entropy, that is present in the food it eats, the water it drinks, the air it breathes, and the energy it absorbs. Here on the earth, the order in the food that man eats ultimately comes from the sun, which supplies the plants with a highly ordered form of energy; without the light of the sun, which is a highly structured form of energy and which plants need for photosynthesis, there would be no food for man to eat, and so, man would soon die unless he could learn how to synthesize food from inorganic molecules, as the plants do. But this, too, could not be done without a highly ordered form of energy such as that from the sun or some other source, such as nuclear energy.

Life and the Laws of Nature

The most striking thing about any single living organism is that it is a unique group of atoms and complex molecules that—contrary to the natural tendency of all inanimate systems to evolve to states of complete disorder and equilibrium—produces orderly events and higher states of order by assimilating inanimate matter and producing from it exact replicas of its own basic entities—its cells. The ability of each living cell to reproduce itself identically from the material it feeds on may, in itself, not appear so remarkable, since something similar seems to happen when an inanimate crystal such as ice or salt grows from a liquid or directly from a gas. Here, too, a small group of atoms or molecules collects other identical atoms or molecules from its surroundings to form a precisely ordered structure that is repeated over and over again; a crystal thus grows by an orderly replication of a definite pattern, but what a difference there is between the exact and endless reproduction of a single pattern in crystals and the hundreds of variegated patterns in living cells. But a more remarkable distinction between a living cell and a large crystal is that the cell reproduces itself over and over again by extracting what it needs from a large variety of atoms and inorganic and organic molecules that are different from itself, whereas the crystal in its growth can use only one kind of atom or molecule; the salt crystal can only incorporate into itself the sodium chloride molecule as it grows. Moreover, the growth of a cell is strictly limited by a precise biological mechanism in contradistinction to a crystal, which can grow without limit by exact repetition. Most remarkable of all is the cooperative existence within the living organism of an incredibly large number of cells of many different kinds that perform in unison the amazingly complex physical, chemical, and biological operations called living. Thus, although the living organism is a highly ordered structure, its orderliness is not the simple periodic order of the crystal: it is a combination of diverse periodicities subtly woven together, and the genes themselves, which guide the entire structure, are complex patterns of periodicities that act as templates for the orderly processes in, and the precise reproduction of, each cell.

Contemplating all of these features of living matter one may wonder whether such matter is governed by the same laws of nature that govern inanimate matter; one may ask how two such different states of matter can obey the same laws. One must simply keep in mind that the behavior of matter is very much a function of the state of its organization and must expect new features and phenomena to appear when the molecules and atoms in one structure are rearranged into another pattern. Whether an atom of carbon is moving about freely in interstellar space, being buffeted violently in the deep interior of a star, vibrating gently inside a diamond, or performing some kind of function in the human brain, it is governed by, and obeying, the same physical laws in each case. There is nothing in the behavior of any atom in a living organism that distinguishes it from the same kind of atom in an inanimate bit of matter. But just as the collection of carbon atoms that is a diamond differs in many remarkable ways from the collection of carbon atoms that is a piece of coal, so does the collection of atoms in living organisms differ from the collection of atoms in dead matter. Life is as much a property of atoms and molecules when appropriately arranged as extreme hardness and great brilliance are properties of carbon atoms when they are properly arranged. Just as the properties of a diamond cannot be detected in individual carbon atoms, so the properties of a living organism cannot be detected in the individual atoms and molecules that constitute it.

Another analogy may serve here to elucidate this point further. If one knew nothing at all about magnetism, the difference in behavior of a magnetized iron bar and one that is not magnetized would be most puzzling and might even appear mystical. The magnetized iron bar would attract and pick up other pieces of iron, whereas the unmagnetized bar would not, and yet the closest chemical scrutiny of the iron atoms in the two bars would reveal no difference between them. How then is it possible for one collection of iron atoms (the magnetized bar) to do what the other, apparently identical, collection of iron atoms cannot do? A knowledge of magnetism answers the question; magnetic properties are inherent in each iron atom, but when iron atoms are arranged randomly, as in the unmagnetized iron bar, the magnetism of one iron atom is offset by that of another, so that no overall gross effect is observed in the bar. By placing this same iron bar in a strong magnetic field or by stroking it with another magnet, the individual iron atoms in the bar are aligned in such a way that their individual magnetic fields reinforce each other and the bar becomes a magnet. Life manifests itself in a somewhat analogous way when atoms and molecules are aligned in structures in such a way that the life-supporting properties of individual molecules reinforce each other. It is known what force aligns the iron atoms and changes an ordinary piece of iron into a magnet and what force changes the carbon atoms from an aggregate called coal to an aggregate called diamond, but the very subtle force that pushes a collection of molecules over the threshold from an inanimate state of being to an animate state is not known. That there seems to be some kind of threshold that separates the living from the nonliving is most clearly apparent in the behavior of various viruses, which behave like large inanimate crystals under one set of circumstances and like live organisms under circumstances that support life processes. In any case, however contrary to natural physical laws life may appear in the present state of knowledge, there is every reason to believe that it will in time be fitted into the pattern of natural laws, just as all other physical phenomena have been fitted into that pattern.

The Building Blocks of Living Organisms

Although the number of cells in an organism ranges from one to billions and each kind of cell has its own highly specialized function, every cell in every living organism has the same basic structure and characteristics; it is a microscopic blob of material called cytoplasm, the main body of the cell, within which is a smaller and quite different blob of material called the nucleus. The function of the nucleus, which consists of long thin threads of matter called chromosomes, is to issue instructions to the cytoplasm and direct it in the performance of the many different tasks and the production of the complex organic substances that are necessary for the development, survival, and reproduction of the organism. Just as the nucleus contains extremely complex chemical structures that, in a sense, consist of all the chemical production plans and blueprints of the cell, the cytoplasm contains its own complex chemical structures, known as enzymes, which follow the hundreds of instructions sent out by the chromosomes and in accordance with these instructions direct all chemical reactions. Each specialized cell has its own specialized set of enzymes, each of which controls one of the innumerable chemical reactions that must go on if the cells are to perform the tasks that sustain the life and direct the growth of the organism.

The chromosomes in the nucleus of the cell consist of long molecular chains called nucleic acids, which in turn are made up of various combinations of four different kinds of individual molecular units called nucleotides; the most important and remarkable of the nucleic acids is deoxyribonucleic acid (DNA). This large and extremely complex molecule, which in the case of the advanced mammals contains billions of individual atoms, not only sets the pattern for the kinds of proteins that each cell is to manufacture but contains the genetic code that is transmitted via cell division from generation to generation. The DNA molecule is thus the control center of the cell as well as its memory and information storage unit. The nucleic acids constitute only a minute portion of the substance in any living organism; in a human being there is only about one teaspoonful of DNA.

The cytoplasm, which constitutes most of the substance of every living cell, consists of long organic molecules called proteins, which in turn consist of combinations of much smaller molecular units called amino acids; there are just 20 amino acids and any one protein differs from another only in the way that their amino acids are linked together. A slight change in this linkage pattern can alter the chemical property of a protein drastically. Thus, the protein oxytocin, a hormone that stimulates the production of milk in an expectant mother's breast when secreted into the blood stream, differs only in the third and eighth linkage of amino acids from the protein vasopressin, a hormone that increases the blood pressure when secreted into the blood. There are two kinds of proteins: the structural proteins, which are used by the cell to construct such things as its walls, bone, muscle, and hair; and the catalytic proteins or enzymes, which monitor the chemical reactions in the cell.

Since all proteins consist of amino acids and all nucleic acids consist of nucleotides, the amino acids and the nucleotides are really the basic molecular units. Although the proteins and the nucleic acids are very long chains of either amino acids or nucleotides, the amino acids and nucleotides themselves are relatively small, although still quite complex, molecules consisting, in each case, of about 30 electrically bound atoms of hydrogen, carbon, oxygen, and nitrogen. Among the 20 amino acids are such acids as glycine, alanine, and cysteine, but there are just 5 different nucleotides: adenine, guanine, cytosine, thymine, and uracil, which may be labeled with the letters A, G, C, T, and U. The nucleic acid DNA, which exists only in the nucleus of the cell, consists of various combinations of only the four nucleotides A, G, C,

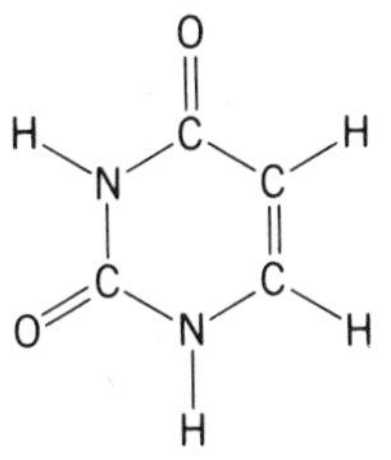

Uracil

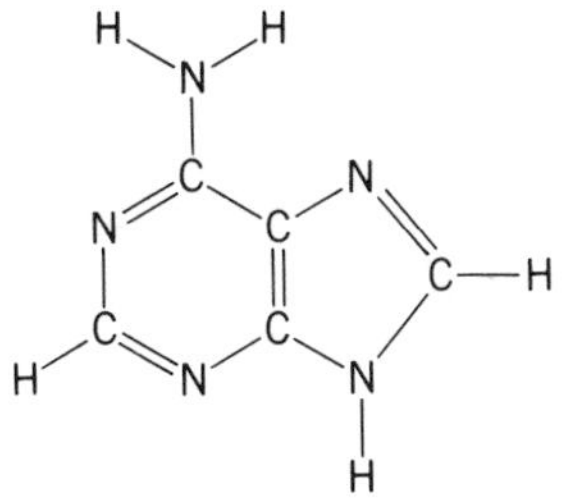

Adenine

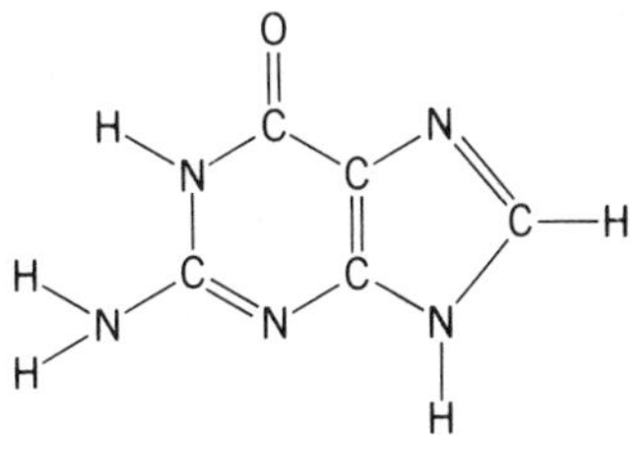

Guanine

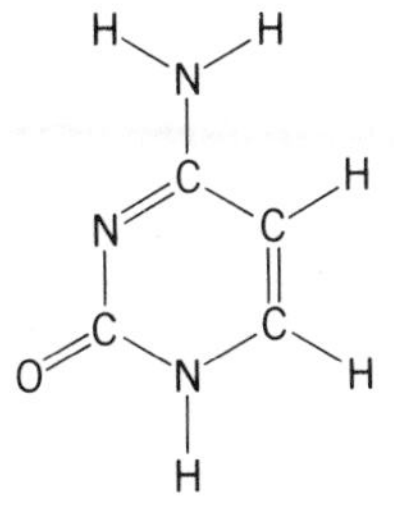

Cytosine

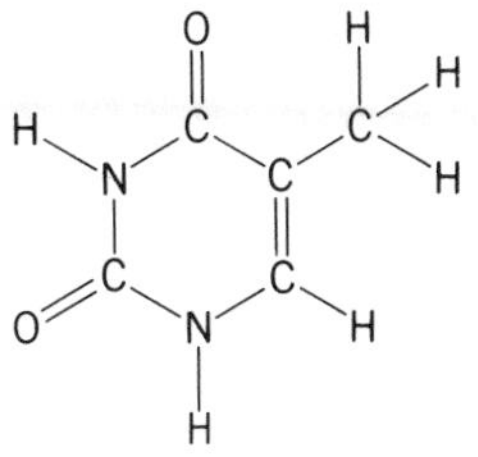

Thymine

The structural formulas of the five nucleotides of protein.

and T, arranged, not in a single strand, but in two long strands that are twisted around each other in the form of a double helix, a structure similar to a spiral staircase in which the rungs are the molecular bonds that bind one strand of DNA to the other. The two strands that form the double helix of any particular DNA molecule are twisted around each other in such a way that each one of the adenine molecules in one strand is bound to only one of the thymine molecules directly opposite it in the other strand and each one of the cytosine molecules in one strand is bound to one, and only one, guanine molecule directly opposite in the other strand. Thus, each rung of the spiral staircase always goes from an A to a T or from a C to a G so that the two strands in the double DNA helix can attach themselves to each other in one, and only one, way. It is clear from this that once a single strand of the double helix is formed from individual nucleotides arranged in any given sequence—and it is this sequence that determines the characteristics and genetic properties of the organism—the second strand of the nucleotides that can attach itself to this first strand is completely determined; if two different nucleotides in such a strand were interchanged, this strand could no longer attach itself to the first strand. To illustrate this, consider a piece of a DNA strand consisting of nine nucleotides arranged in the scheme

A G A C T A A G G,

where the lines indicate the molecular bonds of the nucleotides. The only other strand of nucleotides that can attach itself to this one is the one with its nine nucleotides arranged precisely in the sequence

T C T G A T T C C;

any other arrangement will not work. The double strand obtained is as follows:

If one then pictures such strands as extended lengthwise about a millionfold and twisted around each other like the threads of a double-threaded screw, one obtains a model of DNA. Once a double-helical strand has been formed, the molecular bonds of the nucleotides are saturated and no other nucleotides can attach themselves to the DNA molecules thus formed.

It should be noted that there is nothing in the mechanism of the formation of DNA from the atoms of hydrogen, carbon, nitrogen, and oxygen and their own molecular combinations that cannot be accounted for by the atomic valences and molecular bonds, all consequences of the quantum theory and the electromagnetic force. In other words, the spontaneous arrangement of atoms and molecules into the basic building blocks of life is a direct consequence of the physical laws that govern atoms and molecules.

DNA and the Life Processes

To understand how DNA controls and directs all living processes—in particular the buildup of proteins—in the cell, consider the role of ribonucleic acid (RNA), which is found mostly in the cytoplasm of the cell, and the structure of the 20 amino acids, which are the building blocks of the proteins. RNA differs in its structure from DNA in that the nucleotide uracil is used in place of the nucleotide thymine throughout. RNA is now known to be the messenger that carries the genetic code from DNA in the nucleus of the cell to the site of protein synthesis in the cytoplasm and thus directs the formation from the basic amino acids of the various kinds of proteins that the cell requires. Each protein consists of various combinations of the 20 different amino acids, whereas the RNA is composed of 4 different nucleotides, so that one is presented with the problem of explaining how a structure (the protein) built up from 20 different units (the amino acids) can

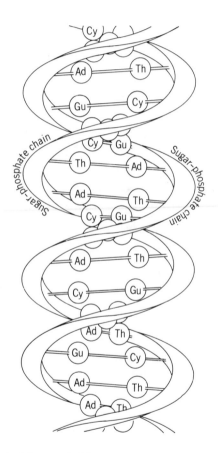

A fragment of the double-stranded helical DNA molecule, with sugar-phosphate chains attached to the outside.

(Right) A fragment of one of the strands of DNA showing how the nucleotides arrange themselves in a chain; the sugar molecules are shown on the right-hand side.

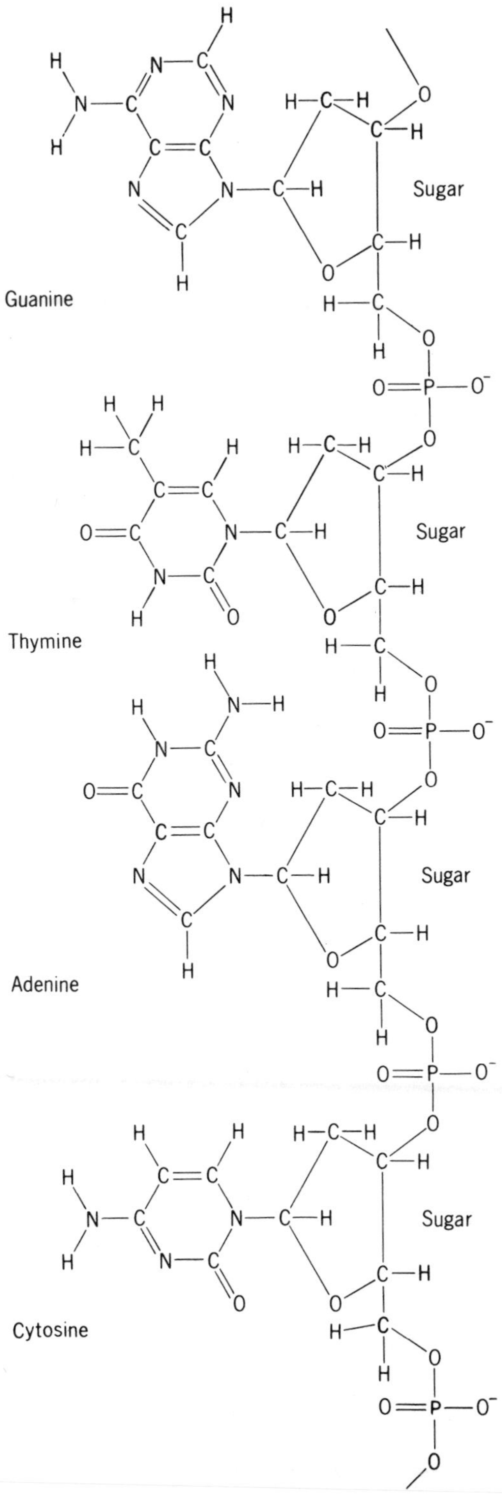

Guanine

Thymine

Adenine

Cytosine

receive its structural information from messengers (RNA) that use combinations of only 4 basic units (the four nucleotides). A great deal of research was devoted to this problem until investigators discovered that each amino acid is defined by a combination of just three nucleotides in RNA. But the number of different combinations of three letters, such as AAA, ACG, AUC, and CUA, that can be obtained if each of these letters can be chosen from the four letters A, C, G, and U is just 64, since each one of the three letters in each triplet can be chosen four different ways. This gives altogether 64 triplets, which is clearly more than are needed to account for only 20 different amino acids. This was puzzling at first, but it was then discovered that some of the amino acids can be formed by two or more nucleotide triplets. Thus, the amino acid proline is defined by any one of the triplets CCC, CCU, CCA, and CCG, whereas the structure of phenylalanine is determined by the triplet UUU only.

The way in which the proteins are built up from the raw material in the cytoplasm of the cell is now fairly clear. The various strands of RNA are assembled in the nucleus of the cell according to the various codes contained in the double-stranded DNA molecule in the nucleus. Once an RNA strand has been constructed from the nucleotides A, C, G, and U in the nucleus in accordance with one or more of the genetic codes on some part of one of the chromosomes, the RNA strand leaves the nucleus and moves into the cytoplasm, stationing itself at some particular site where some particular protein is to be synthesized in accordance with the RNA blueprint. After completing its job at one site it moves to some other site to repeat its construction work.

Not only do the RNA strands direct the construction of the structural proteins but also that of the catalytic proteins that enable chemical reaction to proceed inside a cell at body temperatures and at rates that would ordinarily require hundreds of degrees centigrade. As an example of this, note that the enzyme pepsin can break down proteins in a matter of a few hours inside the stomach at body temperatures, whereas in the laboratory the same proteins have to be boiled for more than 24 hours in a strong hydrochloric acid solution to be broken down. A single living cell may contain as many as 100,000 enzyme molecules to speed up the 1,000 or 2,000 chemical reactions that go on in the cell. Thus, only a few enzyme molecules are required for each chemical reaction in the cell. A single enzyme molecule can transform anywhere from 1,000 to 500,000 other molecules per minute. When hydrogen peroxide is poured on an open wound, a white foam develops because of the presence in the blood of an enzyme that breaks up hydrogen peroxide into oxygen and water, the oxygen being the antiseptic agent that destroys germs; a single molecule of this enzyme dissociates 1 million hydrogen peroxide molecules per minute.

The most remarkable of all the properties of the DNA molecule is its ability to reproduce exact patterns of itself and thus to produce cell division and to propagate the species. The double-stranded structure is absolutely essential for this process, as is the one-to-one bonding of the nucleotides A to T and C to G. The cell division begins with the lengthwise splitting of each DNA molecule in the nucleus of the cell after the cell has reached its full growth. At that time in the history of a cell the rungs of each DNA ladder that are relatively weak, the A—T and C—G molecular bonds, begin to break, starting at one end of the double strand and progressing down to the other end, like the two strands of a zipper that is being opened. While the DNA molecules (the chromosomes) are all splitting in this way, the nucleus of the cell fades away until nothing is left of it but a double set of chromosomes, which then wander to the cell's equator, where they line up lengthwise along the equator. During this process, starting shortly after the splitting of the DNA molecules has occurred, each of the two single strands of each of the DNA molecules captures new nucleotides from the pool of free nucleotides that are present in the nuclear material in exact accordance with the A—T, C—G bonding so that two new identical DNA molecules for each of the old ones are formed. When this doubling of the DNA molecules is complete, so that two identical sets of these molecules are present, one set wanders to one pole of the cell and the other set wanders to the other pole. The cell then splits along the equator into two daughter cells, each of which has the same set of DNA molecules and hence the same genetic information and heredity as the parent cell.

Viruses: Between the Inanimate and the Animate

Although each of the molecular processes that we have discussed in the previous sections and that, taken in toto, constitute life can be explained singly in terms of the electromagnetic forces and the various molecular bonds—particularly those of carbon—that tie molecule to molecule, the explanation of, and the reason for, the essential features of life, which is an extremely complex cooperative phenomenon involving the orderly and yet simultaneous operation of thousands of chemical reactions, escape us. But organisms exist that seem to hold a clue to this important question, for they exhibit both animate and inanimate features. One can picture a number of stray nucleotides, formed from a random association of atoms of hydrogen, carbon, nitrogen, and oxygen in the air or in the waters of some murky swamp, coming together and linking up to form a piece of a DNA strand or even to form part of a double strand, but such a structure would not be a living organism, although it might be on the way to becoming one. The minimum requirements that such associations must fulfill before they can be considered living organisms might be found in the structure of viruses, which are the smallest disease-inducing organisms known.

The word *virus* was introduced into biology and medicine in 1889 by the Dutch botanist and bacteriologist Martinus W.

Beijerinck, who suggested that the "tobacco mosaic disease," an infectious disease of the tobacco plant that mottles the leaves of the plant, is not caused by a germ but by a venom (hence the term *virus,* which is the Latin word for poison). Nobody knew the chemical nature or structure of the tobacco mosaic virus until it was isolated and crystallized in 1935 by the American bacteriologist H. A. Allard. The remarkable thing about this virus in its crystalline phase is that it exhibits none of the properties associated with life and looks very much like any other inanimate crystal; it is as lifeless as particles of salt or sugar, neither growing by the subdivision of individual crystals nor by the ingestion and assimilation of molecules and atoms from its environment. But if these virus crystals are dissolved in water and placed on a tobacco leaf, they immediately begin to live and multiply, spreading from leaf to leaf and forming huge virus colonies rapidly.

With the introduction of the electron microscope, which uses high-speed electrons to "see" objects that are too small to be visible in ordinary microscopes, many other kinds of viruses have been discovered; they come in various sizes, the smallest being about one-millionth of an inch in diameter, and may be circular, cylindrical, spherical, or polyhedral, with some viruses exhibiting tails and looking like tadpoles. Chemists have analyzed these remarkable organisms and have discovered that they consist of a nucleic acid core and a protein skin or shell that completely encases the core. Apparently a virus is a bagful of pure genetic material that comes to life only when it is in contact with a living cell; otherwise, these minute creatures collect themselves into inert, lifeless crystals. The process by which viruses attack living cells is now understood; a single virus attaches itself to the wall of a cell, in accordance with the genetic instructions imprinted on its DNA, and then, via some kind of appendage of its own skin (which is all protein, like the stinger of a wasp) pierces the cell wall and injects all of its DNA into the cell, leaving behind its empty protein coat. At this point the intruding DNA of the virus is attacked by the natural immunological defenses of the cell, which in most cases destroy the foreign DNA. But if these cell defenses are not strong enough to devour the injected DNA molecules, these molecules, using their own genetic code, induce the cell to manufacture new viruses from its own nucleotides and amino acids, and within 10 minutes of the intrusion, new empty protein virus coats are formed inside the cell. Some 10 minutes later the regions inside the new protein coats begin to fill up with viral DNA, and after another 10 minutes—30 minutes after the initial intrusion—the cell wall bursts, releasing into the surrounding medium a few hundred new viruses.

These phenomena indicate that viruses are indeed the link between inanimate and animate matter, but the characteristics of a virus that are called life can be evoked only by something that is already living, so that there is a discontinuity or a threshold that still cannot be explained. There is a kind of triggering action that occurs when a virus comes into contact with a living organism that still has to be explained: the kind of force or electrical impulse that the living cell conveys to the virus that sets the life processes going in the previously nonliving virus. This question still has to be answered, and when it is, an understanding of the origin of life will not be far away.

But even without having an answer to this question one can already see, again by studying the structure of a virus, how life might have begun. In 1955 the American biochemists Heinz Fraenklin-Conrat and Robley Williams extracted the nucleic acid core of the tobacco mosaic virus chemically from its protein skin. Separated from each other in different containers, the viral DNA and the viral protein looked like any other complex organic molecules. Neither the DNA by itself nor the protein showed any signs of life, even when placed on live tobacco leaves. However, when the protein solution and the DNA solution were mixed together, the protein molecules began to organize themselves into a skin around the nucleic acid molecules, and soon complete tobacco mosaic viruses were reconstituted and again arranged themselves into inert crystals when dry. But when these reconstituted viruses in solution were placed on live tobacco leaves, they behaved just like the untouched viruses, and began to infect the cells of the tobacco leaves, multiplying as if nothing had happened to them.

Here, then, is an important clue to how life might have begun from inanimate organic molecules; one need merely postulate that certain types of protein molecules or amino acids and certain types of nucleotides have strong affinities for each other and that when they meet by chance, they combine as would two matching pieces of a jigsaw puzzle. The evidence presented by the structure of viruses makes this more than a postulate and strengthens the belief that viruses were formed piece by piece, over a period of millions of years, through the chance encounter of amino acids and nucleotides. One may assume further that when a nucleotide and an amino acid combined, the chance for the capture of additional nucleotides and amino acids was greatly enhanced and that this went on until a complete virus was formed. If, then, one accepts this picture of the chance formation of viruses, one may accept them as the precursors of living organisms. Imagine the warm waters of a shallow sea about a billion years ago, rich in inanimate organic molecules of all kinds, among which were many varieties of nucleotides and amino acids, suffering random collision. In time, such collisions resulted in various combinations of these two types of organic molecules and they in turn united to form complete viruses, which would still not be living organisms, for viruses cannot live by themselves. It is possible that some types of these very early viruses had affinities for each other and fused together under appropriate circumstances to form living cells. Although this is all guesswork, the model presented here contradicts none of the known laws of physics or chemistry.

One may wonder whether or not new viruses are being

formed constantly in the oceans today. They probably are, but their chances for survival and buildup into complete viruses are extremely small because all organic matter is food for some form of life or other, and life permeates every nook and cranny of the hydrosphere; such rich food would be gobbled up quickly, which did not happen billions of years ago when few living organisms were around to prevent others from arising.

Origin of Nucleotides and Amino Acids

If the formation of viruslike structures from completely inanimate organic material was, indeed, the first step that led to life and its evolution from one-celled creatures to its highest mammalian forms, the problem of the origin of life is really the problem of the spontaneous formation of amino acids and nucleotides from the simple organic compounds and inorganic molecules that existed in the primordial planetary atmospheres. Organic chemists today can synthesize in their laboratories all the amino acids as well as the four nucleotides, but only under highly controlled conditions. But this is a far cry from the spontaneous synthesis of such compounds under natural conditions; nevertheless, there exists conclusive experimental and observational evidence that this kind of organic synthesis does occur. Urey reasoned that the ultraviolet rays from the sun and various kinds of electrical discharges to which the primitive planetary atmospheres were subjected a few billion years ago triggered a chain of chemical reactions that led to the synthesis of nucleotides and amino acids from the methane, ammonia, water vapor, and carbon dioxide that were present in those atmospheres. To check this hypothesis, Urey had one of his graduate students, the American chemist Stanley L. Miller, subject a mixture of hydrogen, ammonia, methane, and water vapor in a test tube to a series of slow electrical discharges for a few days. On analyzing the content of the test tube at the end of this time, Miller found that several of the 20 amino acids had been formed, thus confirming Urey's conjecture.

Since it may be argued that the Miller-Urey experiment depended on special kinds of electrical discharges that might not have occurred in the earth's primitive atmosphere, it is important to note that even stronger evidence for the picture drawn above of the spontaneous production of amino acids exists in the recent chemical analysis of meteorites that were recovered shortly after they had struck the earth. The purpose of this analysis was to determine whether such meteorites contained either amino acids or the complex organic compounds that lead to amino acids. Chemists have been studying meteorites for over a century and everyone now agrees that organic matter is present in and on these objects, but until recently the conclusions about the origin of these organic compounds were ambiguous, for it was not definitely established that the meteorites had not been contaminated by terrestrial organic matter upon striking the earth. But the evidence against this terrestrial origin of the organic material on meteorites is now conclusive, for the Ceylonese-American astronomer Cyril Ponnamperuma has applied special techniques to the chemical analysis of a recent meteorite that differentiate clearly between terrestrial and nonterrestrial organic compounds; his analysis shows that the 19 amino acids that are present in this meteorite were formed outside the earth, either in interplanetary or interstellar space.

The novel technique that Ponnamperuma used depends on the fact that two isomers of each organic compound exist that are exact mirror images of each other: thus, every amino acid has its mirror image as a possible organic compound. The difference between a left-handed organic isomer (levorotatory or L-type) and its right-handed counterpart (dextrorotatory or D-type) can be recognized by noting how such an isomer rotates the plane of polarization of a plane of polarized beam of light; the levoisomers rotate the plane to the left as the polarized beam of light passes through a levoisomer solution, whereas solutions of dextroisomers rotate the plane to the right. By analyzing the rotatory properties of the amino acids found in meteorites Ponnamperuma demonstrated that the levo- and dextroamino acids were present in very nearly equal quantities in the meteorite studied. This means that these amino acids could not have originated on the earth because all organic compounds produced on the earth by living organisms are levorotatory; dextrorotatory organic compounds, such as the dextrosugars, can be synthesized by organic chemists, but living organisms cannot metabolize or assimilate such D-type organic molecules. In fact, when chemists manufacture sugar in their laboratories, such sugar in solution does not rotate the plane of polarization of light at all, showing that equal quantities of L- and D-type were produced. If this artificial sugar is fed to bacteria, only half of the sugar is consumed, and the half that remains rotates the plane of polarization to the right (D-type). This means that if all the amino acids found on meteorites were terrestrial contaminants, they would all be L-type, which is contrary to the observed data. From this one must conclude that amino acids are produced spontaneously in outer space from simpler organic compounds, under the most diverse physical conditions.

Now one can go one step further and see what the situation might have been before amino acids were formed spontaneously either on the earth or in outer space and ask what kind of simpler organic molecules could have been used as building blocks for the amino acids. The answer has been given by organic chemists, who have shown that mixtures of simpler organic compounds, such as formaldehyde, ammonia, and hydrogen cyanide, can be converted into families of amino acids by simple heating. This was first done by the American chemists Sidney W. Fox, K. Harada, G. Krampity, and G. Mueller in 1970, who also showed that this also happens in the presence of lunar dust; these investigators also proved that some types of amino acids are present in the lunar dust itself. The importance of the discovery that groups of amino acids are formed from such compounds as formaldehyde, ammonia, and other hydrocarbons under simple physical conditions that might be found in certain regions of the Galaxy is important because these very compounds and many others have been found in interstellar space. Some 30-odd different complex chemical compounds, ranging in complexity from hydrogen cyanide to acetaldehyde have been identified with the aid of large radio telescopes, which can pick up the radio waves emitted by these molecules; many of the known alcohols have thus been found among these compounds.

More exciting even than this is the discovery that the nucleotides and such extremely complex organic molecules as the porphyrins—the molecules that are basic structures from which chlorophyll molecules and the hemoglobins of human blood are formed—can also be readily synthesized in a random way from the interstellar molecules mentioned above.

Clearly, the origin of life is a natural chemical process that can be understood in terms of the chemical properties of atoms and molecules; and these in turn can be traced back to the electromagnetic forces between electrons and protons. Two other crucial ingredients, however, must be available to the organic molecules if they are to evolve from lifeless things to living organisms: time and energy of the right kind. Fortunately for man both of these were available to terrestrial molecules because of the properties of the sun and the position of the earth in the solar system. The sun and all other stars like it in the Galaxy have just the right properties to allow life to evolve on a planet like the earth, which is neither too close nor too far away from the sun. Since about a billion years are needed for the formation of amino acids, nucleotides, viruses, and living cells, the sun has to be the kind of star that can supply a planet like the earth with the right kind of radiation for a few billion years at least, and the sun is, indeed, the right kind of star. The mass of the sun when it was formed was just large enough to have permitted it to reach its present stage in a billion years or so and to have kept it radiating pretty much as it now is for some three or four billion years since then. Stars more massive than the sun are not favorable for the evolution of life because they are far too luminous, thus burning out too quickly, and they emit far too much ultraviolet radiation. On the other hand, stars not as massive as the sun are too cool to activate the chemical reactions that lead to living organisms. Thus, stars like the sun are the most probable objects around which life-supporting planets will be found, but this leads to the prediction that there are probably hundreds of millions of planets in the Galaxy with living organisms on them. An analysis of the abundances of the various types of stars in the Milky Way shows that there are anywhere from 800 million to 1 billion like the sun, and since a planetary system like the sun's was probably formed in each case when these sunlike stars evolved from dust and gas, there are probably about 800 million to 1 billion life-supporting planets in the Galaxy. Some of these are probably populated by creatures that are far below man's state of development, but there could be many that have civilizations far in advance of man's. Man is certainly not at the bottom of the heap, but neither is he at the top; galactically speaking, man has just entered his adolescence. Considering that the universe contains about 100 billion galaxies, there may be life throughout the universe.

11 The End of the World

A Hierarchy of Endings

So persistent is life that one has difficulty contemplating his own nonexistence, let alone the cessation of all life on the earth; and if it is difficult to conceive of an earth devoid of life, how much more difficult it is, then, to conceive of the end of the world itself and ultimately the end of the universe as it is known. When people in the past thought about such matters, they generally either dismissed them as impenetrable mysteries or accepted a simple answer based on established religious doctrine. The idea that such questions might, in time, be answered in a fairly precise scientific way was never seriously considered until quite recently when the vast development in electronic computers began. Now, with computer technology at a stage where very complicated mathematical problems, involving many variables, can be solved in a matter of minutes, hours, or, at most, days, the future history of such fairly complex physical systems as planets, stars, galaxies, and even the universe can be deduced, if not precisely, at least with sufficient accuracy to permit one to draw a fairly reliable picture of the various ways in which the world might end. But it will be impossible to choose from among these various possibilities. The reason for this uncertainty about the manner of the ultimate destruction is that all the possible endings described here can occur only after billions of years have elapsed, and present knowledge of all the factors that can influence and affect the final state of affairs is hardly accurate enough to permit a definitive single choice of an event that is so far in the future.

In considering the various paths along which the world may move to its final destruction, one can arrange them in a hierarchy in which those listed later include those listed earlier. To a person concerned deeply with human life and nothing else, the disappearance of human life from this planet would mean the end of the world even if other life still remained; and in a sense this is the simplest and least inclusive definition of the end of the world. If all life were to disappear from the earth owing to some catastrophe, human life would also disappear, so this would be a more inclusive, and thus more drastic, "ending" than the cessation of human life alone. More inclusive still, with no possibility of the renewal or reemergence of life here, would be the destruction of the entire planet by some cataclysm or other, but this would be a less drastic ending and less inclusive than the disappearance of, or radical change in, the sun itself, for then the entire solar system, and with it the earth, would either disappear or be drastically altered. Finally, an end to the entire universe or a profound change in it would alter all structures within it and could thus end our world and all other worlds.

The Impact of Man

Much has been said and written about the industrial and military activities of man himself that are dangerous to life, and they are, indeed, proper topics to be considered in an analysis such as this, but most of them can be dismissed in a few words. A nuclear world war would result in immeasurable destruction of life directly and would possibly leave in its wake a poisoned atmosphere that would take its toll of life for many more years after the war, but it is highly doubtful that life in general, or even human life, would be entirely destroyed. A careful analysis of the events following such a holocaust indicates that human life would probably survive such a war, but the destruction would be severe enough to retard progress for decades. In any case, since a nuclear war is not a natural event—except insofar as we may treat the deliberate decisions and action of men as natural because man himself is a product of nature—it will not be considered any further here.

There are other slow nonnatural processes that stem from highly industrialized society, and they can in time affect life drastically or even lead to its ultimate destruction. Among them are the processes that introduce an ever-increasing concentration of contaminants into the atmosphere, which can alter the balance between the rate of absorption of visible solar radiation by the earth's surface and the rate at which the earth emits infrared and heat rays. Owing to increasing industrial activity, there is a slow increase in the abundance of carbon dioxide and other infrared-absorbing molecules in the atmosphere. As the concentrations of these gases in the atmosphere increase, the resulting increased greenhouse effect causes the surface temperature of the earth to rise; in time, then, if this effect is not controlled, it could become large enough to destroy life or alter it considerably. At the present time the contribution of industrial activity to the concentration of carbon dioxide in the atmosphere is negligible compared to the contribution by volcanoes and other natural phenomena, and plant metabolism is quite able to maintain a healthy balance between atmospheric carbon dioxide and atmospheric oxygen.

Because the carbon dioxide–oxygen balance in the atmosphere is controlled by plant life, any profound change in the abundance or distribution of plant life would bring with it an equally profound change in the atmospheric abundance of carbon dioxide and free oxygen; this, in turn, would certainly affect human life. Man must therefore be concerned about the steady encroachment that he is making on the forests and the oceans. It has so far had only a slight effect on the abundance of free oxygen, but a critical point will undoubtedly be reached at some future time and some action will then have to be taken to prevent any further depletion of atmospheric oxygen. It should be noted in this connection that if plant metabolism did not renew the supply of free oxygen in the atmosphere, it would, in about 3,000 years, combine with the chemicals in the earth's crust and disappear. This means that the plants emit, via photosynthesis, as much free oxygen every 3,000 years as is present in the atmosphere at any moment. Plant life is just as important for regulating the abundance of atmospheric carbon dioxide as it is for replenishing the free atmospheric oxygen. Since photosynthesis uses the carbon dioxide that is ejected into the atmosphere by volcanic activity, there is no danger that too much carbon dioxide will accumulate and thus lead to a runaway greenhouse effect. It has been estimated that plants consume 500 billion tons of carbon dioxide every year, transforming it and water into free oxygen and organic matter. This process, of which 90 percent is carried on by the microscopic green algae in oceans and 10 percent by the green land plants, combines about 150 billion tons of carbon with about 25 billion tons of hydrogen from the water to form complex organic molecules and releases 400 billion tons of oxygen. Since the total mass of carbon dioxide in the atmosphere at any moment is about 3.5 trillion tons, all the atmospheric carbon dioxide would be used up in about seven years if volcanic activity and, to some extent, animal metabolism and the decay of vegetation did not replenish the supply.

Man's industrial activities may disturb the ozone layer, which is at a height of about 50 kilometers above the earth's surface and which plays an important role not only in the oxygen-water balance in the atmosphere but in the development of life itself. Ozone is the triple oxygen molecule O_3, which readily absorbs ultraviolet radiation; without the thin layer of ozone, whose ozone content is quite small, the harmful ultraviolet radiation from the sun would destroy human life and possibly all of nonmarine life in a very short time. Moreover, this same ultraviolet radiation would disrupt water molecules, releasing free hydrogen and oxygen. The free hydrogen would escape from the earth, and the oxygen would oxidize matter in the earth's crust. Thus, the earth's water would gradually disappear if the protective ozone layer were destroyed, which some atmospheric scientists fear may in time happen if there is a large increase in the number of flights of jet airplanes. But it is unlikely that these dire predictions will be fulfilled, because mankind will, as the need arises, control and check any of its activities that become too injurious to life.

Natural Terrestrial Catastrophes

There are natural events that may destroy life or alter it drastically. Geological evidence shows that great changes have occurred on the earth resulting in vast inundations of continents by water at certain periods in the past and the formation of glaciers during the great ice ages that covered up to 75 percent of the North American continent, half of Europe, and all of Siberia. The inundations by water occur between periods of mountain-building, whereas glacier building occurs during periods of mountain formation. The enormous ice sheet that now covers most of Greenland, with an area of more than 700,000 square miles and an average thickness of about a mile, is the residue of the last glaciation period in the Pleistocene epoch; if this ice sheet were to melt, the level of all the oceans in the world would rise about 30 feet and most of the lowland countries such as Holland would be inundated. A much larger permanent ice sheet covers the continent of Antarctica; it is more than a mile thick and covers an area of about 5 million square miles. If it were to melt suddenly, most of the earth's land would be covered with about 200 feet of water. This, of course, would not destroy life on the earth but would alter it drastically. There appears to be very little likelihood that the thermal conditions on the earth could change drastically enough to cause the various glaciers to melt suddenly, so one can dismiss that as a serious threat to life. But there is a very slow annual decrease in the sizes of the earth's glaciers that causes the ocean levels to rise by about one inch per year, but such changes do not go on in one direction steadily; they alternate with periods during which glaciers are built up. It should be noted that to melt all the ice and snow covering the polar caps would require as much energy as earth receives from the sun in 2.5 years. Hence, thousands of years would be needed to melt all the glaciers even if the glaciers went on melting each year at the present rate instead of alternating between melting and freezing.

Other events that can have drastic effect on life are those that lead to the periodic occurrences of ice ages and warm periods; geological evidence indicates that during the last 600,000 years there have been periodic advances and retreats of ice sheets similar to the one mentioned above. The last ice age, which occurred about 25,000 years ago and extended as far south as Kansas City in the United States, was preceded some 40,000 years ago by a tropical period in the northern hemisphere. Numerous fossils of palms and other tropical plants have been uncovered in the deposits of northern Europe; fossils indicate that Greenland, Alaska, and northern Norway were quite warm and were covered with heavy forests of oak, maple, and birch. But 60,000 years ago the world was in the grip of an ice age, and some 150 million years ago periods of glaciation alternated with periods of warmth.

Although many factors enter into the long-range temperature and climatic changes that occur on the earth, a few factors stand out as more important than others. Thus, there is fairly strong evidence that long-range changes that lead to long glaciation periods are associated with mountain-building over millions of years on the earth, but there are relatively short frigid periods within these long periods, which are probably regulated by the slight periodic changes that occur in the tilt of the earth's axis and the shape of the earth's orbit. Because the earth's orbit is an ellipse, its distance from the sun changes from month to month; at the end of December, when the earth is closest to the sun, its distance from the sun is about

91.5 million miles, but in June, when it is farthest from the sun, its distance is about 94.5 million miles. Owing to the variation in the earth's distance from the sun, the winters in the northern hemisphere are warmer than those in the southern hemisphere and the northern summers are cooler than the southern ones. This effect is not very pronounced now, amounting to a difference of only some 8°F. But if the northern summers were to become about 20°F cooler during a period of mountain-building, the ice formed from the snow deposited during the winter months would not melt entirely; the resulting accumulations from summer to summer would then lead to an ice age. Right now this cannot happen because the northern summers are still too warm, but the gravitational perturbations of other planets, particularly Jupiter, acting on the earth's orbit are slowly elongating the orbit; periodic elongation of the orbit alternates with periodic rounding of the orbit. In time, then, the difference in the earth's distances from the sun when the earth is closest and farthest from the sun can become as large as 11 percent. If the northern summers then occur when the earth is farthest from the sun, they will be cool enough for an ice age to set in.

This tendency toward cooler summers, stemming from a gradual elongation of the orbit, is increased by a gradual decrease in the tilt of the earth's axis. The reason that seasons change on the earth as it revolves around the sun is that the earth's axis of rotation is tilted 23.5° with respect to a line perpendicular to the plane of the earth's orbit, so that the northern and southern hemispheres alternate in pointing toward the sun. In June the north pole is tilted toward the sun, so that summer then begins in the northern hemisphere; the northern winter begins at the end of December, when the north pole is tilted away from the sun. If this tilt were smaller, the northern summers would be cooler and this would promote an ice age. It has been shown that the perturbing action of other planets on the earth tends to decrease the tilt of the earth's axis slowly but steadily for a time and then to increase it. The actual change is not very great, but it may be just sufficient after a given period of time to push the earth into an ice age. If some 20,000 years from now, when mountain-building is at its maximum, the elongation of the earth's orbit is also at its maximum and the tilt of the earth's axis is at its minimum, an ice age will be started if northern summers occur when the earth is at its greatest distance from the sun. Such a catastrophe will not destroy all human life, but it will lead to great hardships and alter the course of civilization drastically unless man discovers by that time how to prevent the occurrence of an ice age.

Collisions with Astronomical Bodies

There is a balance—or very nearly so—of forces on the earth that does not let any one destructive trend get out of hand and result in a runaway catastrophe. Mankind thus appears to be safe from destruction by any of the natural forces that determine long-range geological processes, but men must consider catastrophic astronomical events beyond the control of any kind of regulatory forces. The kind of event that immediately comes to mind here is a collision of the earth with a very massive body such as a large meteorite, an asteroid, or a comet. The bombardment of the earth by meteorites is, of course, a well-known phenomenon, and such collisions can be quite destructive; but the chance that a meteorite massive enough to destroy the earth entirely, to alter its orbit appreciably, or even to destroy much of the life on the earth will strike the earth is so small that it can be completely disregarded. One need only consider some of the data and statistics of meteorites that have hit the earth in the past. Neglecting those that are too small to be observed, and hence too small to do much damage, one finds that about 24 fairly massive meteorites strike the earth every year, but only about 6 of them hit the land areas. Meteorites in most cases are not single objects but generally consist of many (sometimes thousands of) pieces when they strike the earth; the meteorite that struck Pulutsk in 1869 consisted of more than 100,000 single pieces, most of them small. Of the more than 2,000 meteorites that have been collected since 1900 very few have masses larger than a ton; the two largest known meteorites weigh more than 36 tons. The destructive power of a large meteorite can be estimated by multiplying half the mass of the meteorite by the square of its speed when it hits the earth; this is its kinetic energy. If the speed of a meteorite having a mass of about 100 tons were about 20 miles per second, or about the speed of the earth in its orbit around the sun, just as it hit the earth, its kinetic energy at the moment of impact would be about 13 trillion calories. If all this energy were transformed into heat, it would vaporize 25,000 tons of water, which is not very much.

Although most meteorites are not very destructive, occasionally very massive, destructive meteorites do strike the earth. There are two famous examples of such meteorites: the one that in prehistoric times struck northeastern Arizona, near Cañon Diablo, and left a crater a half a mile across and a thousand feet deep, and the Yenisei Valley meteorite that fell in Siberia in 1908 and destroyed all trees and animals within a radius of 50 miles. Such meteorites weigh more than a million tons and are so destructive that they would immediately and completely destroy any city they struck. But they are so rare (perhaps two per century) and the total surface area of the earth is so large that the chance that a meteorite of this sort will hit a region the size of New York City in a given year is 1 in about 50 million. Thus, man appears to be fairly safe from destruction by meteorites. Aside from the minor destruction in a small area produced by these massive meteorites, they have no effect whatsoever on the earth; neither the earth's rotation nor its revolution around the sun is altered in any measurable way by such a collision.

A collision with a comet, which is considerably more massive than the most massive meteorite, may be a different story. That such collisions are not impossible and have probably occurred many times during the past few billion years can be deduced from general statistical considerations. Consider a comet that comes as close to the sun as the earth is; such a comet has one chance in 40 million of hitting the earth because that is how much larger the surface of a sphere having a radius of one astronomical unit is than the surface of the earth is. Since there are about five such comets in the solar system every year, the earth should collide with a comet on the average of once every 80 million years. If these same statistics apply to the past, the earth suffered about 70 cometary collisions during the 5 billion years of its life. A careful analysis of sedimentary rock deposits and the remains of fossils laid down during geological time indicates that such collisions have had very little effect on the orderly astronomical, geological, and biological processes of the earth. Since the mass of a comet is very large compared to that of a meteorite—but less than one two-billionth of the earth's mass, although Halley's comet, with a tail containing millions of tons of matter, may have had a mass one-millionth of that of earth—a direct collision would be very destructive if all the comet's mass were concentrated in its nucleus. Thus, if the nucleus consisted of bodies that weighed tons, and there would then be many such bodies, a collision would destroy all life over a large region of the earth's surface, but life in general would remain intact. But man's knowledge of the structure of comets when they come close to the sun indicates that they consist of small fragments, most of which would be vaporized by the earth's atmosphere; there would thus be a magnificent shower of shooting stars but very little destruction. The poisonous cometary gases would contaminate the earth's atmosphere to some extent, but probably no more than do the exhausts from automobiles and from industrial plants.

If a comet with a mass one-billionth that of the earth were to hit the sun, which is a very unlikely event, since it would have to be moving almost directly toward the center of the sun all along its orbit, one would hardly be aware of it. If it came from a very great distance, its speed on hitting the sun would be about 400 miles a second and its kinetic energy would be about equal to the energy radiated by the sun in three seconds. If the mass of the comet were one-millionth of the earth's mass, like Halley's comet, it would generate about as much heat on striking the sun as the sun radiates in one hour. And if all of this happened instantaneously, the earth would get unbearably hot; an hour's worth of sunlight poured onto the earth in a second would probably make the lakes, rivers, and oceans boil. But the comet's kinetic energy would not be released instantaneously as heat near the surface of the sun. Just as a person who dives into water slows down gradually as he penetrates deep into the water, so a comet would penetrate quite deep into the sun before giving up all its kinetic energy, which would then be added to the total store of solar energy and released gradually. There might be a very brilliant flash but nothing more. If comets that came within one astronomical unit of the sun moved in completely random directions, there should have been about two comet-solar collisions per century, but no such collision has ever been observed.

The possibility of the earth's colliding with another planet is practically nil. If the planets were moving around the sun randomly there might be some chance for such a collision, but that is not the case at all. Since the size and shape of a planet's orbit are rigidly controlled by two basic dynamical principles that can be deduced from the immutable laws of dynamics and gravitation, it is clear that once a planet is in a given orbit, it just cannot leave that orbit and wander as it pleases around the solar system, for it would then violate these two principles. The principle of conservation of energy governs the size of a planet's orbit and the principle of conservation of angular momentum governs the shape; this means that the size of a planet's orbit can be changed only if the planet receives or gives up some of its energy and that the shape of the orbit can be changed only if the planet increases or decreases its store of angular momentum. In either case the action of some external agent that can exert at least as large a gravitational pull on the various planets as the sun does—for example, the intrusion into the solar system of some other star—would be required to change the orbit of any planet, or even of the asteroids, sufficiently to bring it close enough to any other planet for a collision to occur. Thus, the earth would have to be supplied with about 66 percent more energy than it now has for it to move out to where Mars is and to come close enough to Mars to collide with it; Mars would have to be robbed of two-thirds of its total energy to change from its present orbit to one that lies as close to the sun as the earth's orbit does. The same general considerations apply to a collision between the earth and Venus; either Venus would have to increase its energy by 40 percent or the earth would have to lose 30 percent of its energy for these two planets to get close enough to collide. Such vast changes in the energy of a planet cannot occur through the gravitational interactions of the planets themselves or through the interaction of the planets and the sun; the sun, by means of its large gravitational pull, keeps the planets moving in their appointed orbits. It has been shown, in fact, from the laws of celestial mechanics, that the solar system is extremely stable and that each planet will keep moving in its orbit almost indefinitely. In other words, it is safe to conclude that perturbations of planetary orbits arising within the solar system itself will not change the planetary pattern for at least a few billion years. Clearly, then, to effect any appreciable change in either the size or the shape of a planet's orbit requires the presence in the solar system of a body that is as massive as the sun—in other words, another star.

The chances that some other star will collide with the sun or come close enough to earth's part of the solar system to wreak havoc with the planetary orbits are easy to deduce from the distribution of stars in this part of the Galaxy. The stars are separated from each other by such distances that even if they were moving at random, the chance for a collision would be negligible; a star like the sun would collide with one of its neighbors only once every few hundred billion years. But even this estimate of the chance of a collision between the solar system and another star is too large because the stars are not moving at random: they move in precise orbits around the center of the Galaxy just the way the planets in the solar system move around the sun. From this it follows that there is practically zero chance of a collision between the sun and another star, because each star has a definite amount of energy and angular momentum, as determined by its galactic orbit, which it cannot change.

If the solar system in its orbit around the center of the Galaxy (it revolves in this orbit once every 250 million years) were to encounter a large and fairly dense cloud of dust and gas, the friction between the earth and the particles of dust and gas could slowly rob the earth of its mechanical energy and cause it to move in a smaller orbit. If the earth lost half its present store of mechanical energy in this way, which would take many millions or even billions of years, depending on the density of the cloud, the earth's orbit would become half as big and the earth's mean distance from the sun would be about 46.5 million miles. The earth would then receive from the sun every second four times as much light and heat as it does now, so that the thermal conditions on the surface of the earth would probably be quite unsuitable for life. The length of the year would then be about 130 days and the seasons would change every 43 days. It should be noted that the earth's orbit around the sun now departs from a straight line only by a minute amount according to earthly standards; for every 18.5 miles that the earth moves in a forward direction in its orbit, it is deflected a ninth of an inch toward the sun by the sun's gravitational pull. Although that ninth of an inch may appear very tiny and inconsequential, it plays a very crucial role in man's life and was extremely important for the evolution of life on the earth; it is not to be tampered with if geological processes and the evolution of living organisms are to proceed in an orderly way. Anything like a cloud of dust that hindered the earth's motion around the sun would change that ninth of an inch to a larger number such as an eighth or a seventh of an inch, and bring us too close to the sun for comfort; any agency that decreased that ninth of an inch to a tenth or an eleventh of an inch would take earth too far away from the sun and bring on an ice age.

The chance, however, that the solar system will encounter one or more clouds of dust in its 250-million-year trip around the center of the Galaxy is practically zero; although vast dust clouds are fairly common in earth's part of the Galaxy, they are moving in their own orbits around the galactic center, just as the solar system is, and such orbits do not cross.

The catastrophic events considered up to this point are not characterized by inevitability; those discussed below are more nearly inevitable. Just which one will doom life on earth to nonexistence it is impossible to tell, if for no other reason than that the time scales of such events are so immense.

Destruction of the Earth-Moon System

When observing the endless ebb and flood of the tides, one is aware of some vast interplay of forces in nature that nothing can stop. Even the mightiest tidal waves appear to be far too weak to affect the earth or the moon in any appreciable way. And yet this sometimes gentle, sometimes violent breaking of waves against the shores does indeed affect the earth, and in such a way that ultimately the earth-moon system, as we now know it, will be destroyed.

Anyone who has lived at the seashore or spent a few days there watching the ocean knows that successive high tides occur about 12.5 hours apart. The two high tides that occur at about the same time on successive days are, on the average, separated by 24 hours and 51 minutes. Since the average interval of time between two successive risings of the moon is also 24 hours and 51 minutes, one infers, quite correctly, that the tides and the moon are closely related, as was first pointed out by Newton, who developed a simple, but correct, theory of the tides based on the moon's gravitational attraction. Although the moon is largely responsible for the tides, the sun also contributes; the sun's contribution to the tides is slightly less than half the moon's contribution even though the gravitational pull of the sun on each particle of water is about 180 times greater than that of the moon. The reason the tide-raising ability of the moon is about twice that of the sun in spite of its much weaker gravitational pull is that it is not the total gravitational pull that counts in raising tides; it is the difference between the pull on a particle of water on the side of the earth facing the moon and on a particle on the other side of the earth. This difference is twice as pronounced for the moon as for the sun because the moon is so much closer to the earth. To see this, note that a one gram particle of water on the earth directly beneath the moon, being closer to the moon than the center of the earth, is pulled more strongly toward the moon than a one gram particle of matter at the center of the earth. Thus, the oceans directly beneath the moon are pulled away from the earth as a whole, thereby forming a high tide. On the other hand, a similar particle of water on the other side of the earth, being farther away from the moon than the earth's center, is pulled less strongly than a similar particle at the center of the earth. The earth as a whole is thus pulled away from the water on the distant side of the earth causing this water to collect into another high tide.

As the earth rotates, bringing different parts of the oceans and the continents under the moon, the tides move around the oceans like a long, low wave with one of its crests beneath the moon and the other on the other side of the earth. A person standing on an island watching the moon rise observes the water rising on the eastern side of the island first, then on the southern shore, and, finally, on the western shore; as the moon sets, the water recedes from the western shore. In the deep oceans the tides are essentially a surface phenomenon, so that there is very little friction between the water and the ocean bottom but along the continental shorelines there is considerable friction as the water rushes over the dry land and then recedes. This friction converts the mechanical energy of the water into heat. But since the tides obtain their kinetic energy from the earth's rotational energy, it follows that the constant flow of the tides is decreasing the rate of rotation of the earth and thus lengthening the day. In a sense, the tides are acting like a huge brake on the earth, slowing the earth down by friction, just the way the brakes in a car slow down the wheels.

Although the tidal slowing down of the earth is extremely small and the lengthening of the day is very gradual, it can be measured with great accuracy by comparing the times of occurence of eclipses in recent years with those recorded in antiquity. There is a marked discrepancy between the two if one assumes that the earth has always been a steady timepiece—that is, that the rate of rotation of the earth has not changed during the last 4,000 years. From the discrepancy between the observations and the theoretical predictions of eclipses, one deduces that the rate of rotation of the earth has been decreasing and that the length of the day is increasing now at the rate of about a thousandth of a second per century. The rate at which the rotational energy of the earth is dissipated by the tides in this way is more than 2 billion horsepower, or very nearly 2 billion kilowatts.

The importance of this for the future of the earth-moon system is that the rotational angular momentum that the earth loses as its rotation slows down is transferred to the moon's orbital motion; in slowing down the earth's rotation by dragging the ocean waters, the moon gains the angular momentum lost by the earth and is thus propelled forward in its orbit. This results in a steady recession of the moon from the earth, a steady increase in the size of the moon's orbit, and a steady increase in the length of the month. The rate at which this is happening and the general dynamical features of the earth-moon system in both the past and the future can be computed from the rate at which the length of the day is now increasing. About 4 billion years ago the moon was about 10,000 miles away from the center of the earth, the earth rotated once every 5 hours, and the month was slightly longer than one day. The present length of the day and the month and the present structure of the earth-moon system are the result of tidal action during the last 4 billion years, and the earth-moon system will go on changing slowly in this way until the length of the month and the length of the day are both equal to 47 of earth's present 24-hour days. When that happens, the earth will always keep the same face toward the moon, so that the moon will be continuously visible from one side of the earth only, neither rising nor setting. But this will not happen for many billions of years because of the very slow frictional action of the tides.

Although the moon will then no longer cause the tides to rise and fall, producing, instead, an unchanging double bulge of the oceans, the sun will, and this solar action will slow the earth's rotation still further until the day becomes longer than the month. At the same time the moon will start approaching the earth again, while the earth and the moon together recede from the sun. Thus, the month will get shorter, while the day and the year will get longer. This will continue until the moon's distance from the earth's center is less than 10,000 miles, which is a critical distance for the moon, known as the Roche limit; at that distance the tidal action of the earth on the moon will tear the moon into pieces that will form a ring around the earth. There is strong evidence that the rings around Saturn were formed in this way when one of Saturn's moons came too close to that planet.

Before the moon is destroyed by the earth's tidal action, the moon itself will raise huge tides on the earth and alter the earth's structure enormously. Since the tide-raising force varies inversely as the cube of the moon's distance from the center of the earth, it increases very rapidly as the moon approaches the earth. At a distance of 10,000 miles, the moon would be about 24 times as close to the earth as it is now, so that its tidal action would be about 15,000 times as great. The ocean tides at their maximum on the earth would then be hundreds of feet high and would completely inundate all the land masses in their path as they followed the rising moon. But this would not be the worst of it, for the moon would distort the entire earth by producing large tidal waves within the earth's rocky crust and in the underlying regions. These structural tidal waves, rushing through the earth's interior would set off vast earthquakes and volcanic eruptions. Although the earth itself would not be destroyed by such cataclysms, all land life probably would be.

In time, after the earth had torn the moon into pieces, the violent eruptions and tremors on the earth would cease, the oceans would recede and life would probably begin to develop again on the dry land. But this would still not be the end of the "tidal evolution" of the earth, for the length of the day would continue to increase because of the sun's tidal action, until it became as long as the year, which would then be a few weeks longer than it is now. The earth would then present the same face to the sun at all times, so that one half of the earth would be in perpetual darkness. The side of the earth facing the sun would become an unbearably hot desert and the dark side would be covered with vast sheets of ice thousands of feet thick. These two forbidding hemispheres would be separated by a narrow zone (perhaps a few hundred miles wide) where intelligent life could exist. But the full series of these events will probably never occur, because the time involved is so great that the sun itself will have changed drastically long before the lengths of the day and the year become equal.

The Sun as a Red Giant

The theory of stellar evolution asserts that the sun will evolve slowly from being a typical medium-sized yellow star to being a cool red giant whose luminosity will be thousands of times greater than that of the present sun. From this red-giant stage it will then move rapidly toward its demise, either bursting apart in a single violent explosion as a supernova or suffering a series of minor outbursts. In either case, the consequences for the earth and the other planets will ultimately be the same—incineration and vaporization by the extremely hot gases and radiation emitted by the exploding sun. Since the sun, governed by immutable natural laws, cannot escape this violent death, the solar system, and with it the earth, must ultimately vanish also, regardless of whether the events described earlier occur or not. It is interesting, then, to trace the future evolution of the sun and to correlate events in its life with changes on the earth.

About 5 billion years ago, when the sun was about 4 percent smaller, 38 percent less luminous, and 10 percent cooler at its surface, it was a chemically homogeneous star whose central temperature of about 10 million degrees was hot enough to fuse protons into helium at a steady rate and keep it from collapsing further. Since then, the sun has become chemically inhomogeneous and begun to evolve away from the main sequence and undergo changes preliminary to its ascent into the red-giant branch of the H-R diagram. Its evolution right now is proceding very slowly: it will continue as a main-sequence star, or very nearly one, for another 2 billion years, but by that time it will have exhausted most of the hydrogen in its core, and its structure will then be significantly influenced by the difference between the chemical composition of its core and that of its outer regions. The sun will be a larger, hotter, bluer, and more luminous star at the end of the next 2 billion years than it is now; its surface will attain a temperature of about 6,400°K, and it will be 15 percent larger and twice as luminous as it is now.

The effect of these solar changes on living organisms will be drastic; life will be destroyed unless man's technology advances to a point where effective countermeasures can be adopted. As the sun gets larger and hotter during the coming eons, the intensity of the ultraviolet radiation reaching the earth will begin to increase, and that in itself could destroy life. To offset this, scientists would probably develop procedures for increasing the density and thickness of the ozone layer in earth's atmosphere, which now absorbs most of the ultraviolet radiation sent to earth by the sun. In time, however, the increasing luminosity of the sun will become a greater threat to life on the earth than the increasing ultraviolet radiation. When the luminosity of the sun has increased to about twice its present value, the overall temperature of the earth will increase by about 25 percent. During the hottest days of the summer months in the northern and southern hemispheres and throughout the year in tropical regions the temperature will reach or exceed the boiling point of water; lakes and rivers will dry up or boil away, and even the oceans will boil in some areas. The pressure of the atmosphere will increase by about 25 percent and dense clouds of water will surround much of the desertlike earth. All this will happen unless scientists and engineers devise ways of shielding the earth against the greatly intensified solar rays. One possibility will be to send into orbit around the earth large reflectors that will reflect the solar rays back into space. In addition to reflectors, absorbing surfaces could also be orbited to capture the solar radiation and harness it for useful purposes. In fact, by an appropriate arrangement of reflectors and absorbers the amount of radiation from the sun reaching the earth could be carefully controlled so that the earth's climate itself would be controlled. At the same time, vast amounts of solar energy would be harnessed.

In this way conditions on the earth will probably be

stablized for another 3.5–4 billion years, but by then the sun will begin to change quite drastically and rapidly. It will have become cooler—4,800°—and redder but about three times as large and three times as luminous as now. The earth's temperature will increase to a point where all the oceans will begin to boil. Conditions will remain this way for a few hundred million years, while the sun consumes the last bit of hydrogen in its core, but then, as its pure-helium core begins to contract rapidly, releasing great quantities of gravitational energy, the sun will expand quite rapidly, cooling off and becoming much redder and much more luminous in the process. It will be on its way to becoming a red giant, with its radius increasing steadily until its red hot surface reaches first Mercury, then Venus, and finally the earth. Each of these planets will in turn be vaporized as the solar gases engulf it, and each such puff of vapor will be added to the expanding solar material.

When the sun has reached the giant stage, its luminosity will be a few thousand times greater than it is now and its central temperature will be about 100 million degrees. This will be high enough to trigger the triple helium reaction, and the sun will be on the last lap of its evolutionary trip to a highly condensed state—a white dwarf; a neutron star, or pulsar; or a black hole. The exact details of the way the sun will evolve from its red-giant state to one of these highly compressed final states have not been fully worked out, but it is known that before it reaches one of these possible end states of its existence, the sun will have to pass through a phase during which it loses a certain amount of mass, and whether it does this very explosively or not will determine how condensed its last stages of life will be.

The Ultimate Energy Crisis?

Although the earth can in principle be shielded by means of technology from the intense solar radiation after the sun has left the main sequence but before it becomes a red giant, it cannot escape destruction when the sun becomes a red giant unless it moves so far away from its present orbit that the expanding hot solar gases will not extend far enough out from the sun to engulf it. This is possible in principle because all that the earth requires in order to do this is additional orbital energy. If, in some manner or other, the earth were speeded up in its orbit, it would move away from the sun and revolve in a larger orbit. But if one tried to increase the size of the earth's orbit to the desired value simply by imparting more speed to the earth, one would probably run into extremely difficult technological problems. One could, however, ease these problems somewhat and eliminate some of the difficulties by using the gravitational fields of Mars, Jupiter, and Saturn to help the earth recede from the sun. But even with the help of these planets one would still have to start things off by increasing the earth's orbital energy until the earth got close enough to these outer planets to be able to use their gravitational fields to acquire additional energy.

To see what is involved here, note that if the earth were moving in its orbit at a speed of about 27.5 miles per second, it would escape from the solar system entirely and move off into interstellar space. Its speed in its orbit right now is about 18.5 miles per second, so that the earth would have to be speeded up another 9 miles per second to launch it out of the solar system and away from an expanding sun. The energy required to do this equals the total energy emitted by the present-day sun in about three months, which is a vast and forbidding amount if it all had to be made available in a short time. But that, fortunately, will not be the case billions of years hence when mankind is faced with a threatening sun; long before the sun begins its destructive expansion into gianthood, scientists and engineers, knowing what will be in store for the earth if protective action is not taken, will begin to ease the earth out of its orbit by increasing its speed gradually until it acquires a speed of 27.5 miles per second. If this is done over a period of about 1.5 million years, which is a very short time compared to the time the sun will spend in its pregiant phase, the speed of the earth need not be increased by more than 1 centimeter per second every year. The energy required for this per year is about the same as that emitted by the sun in five seconds. Although this is still a vast amount of energy, it will not be unmanageable for a civilization whose technology will have advanced a few billion years. But if this still proves to be too large an expenditure of energy in the allotted time, the process of speeding up the earth to the required speed can be stretched out over a period of 100 million years, adding a hundredth of a centimeter per second every year. To do this mankind would have to be advanced enough to liberate explosively every second, on the side of the earth opposite to the direction in which it is moving, an amount of energy equal to that liberated by a 15,000-pound hydrogen bomb. Such explosions would propel the earth forward in its orbit sufficiently to increase the earth's speed by the required amount. It would not be necessary to keep this up for the entire 100 million years if the earth could be maneuvered in such a way as to bring it just close enough to Mars to take advantage of Mars' gravitational attraction and then close enough to Jupiter to ride Jupiter's gravitational coattails.

Although this escape from the solar system seems to involve a possible, but extremely unrealistic, energy-expenditure technique, it is clear that this is precisely what will have to be done if the earth is not to be entirely destroyed. It may, indeed, be necessary to transform the earth into a vast spaceship and either set it moving again in some distant orbit around the expanding sun or detach it entirely from the sun and move it into an orbit around some other congenial star not too far from the solar system. If the earth were to remain in the solar system after the sun had become a few thousand times more luminous than it is now, it would have to be placed in an orbit 40 or 50 times as long as its present orbit to survive the intense solar rays. A year on the earth would then equal about 350 of its present years.

If, instead of remaining in the solar system, the earth were to move off into space looking for another sun, the earthly inhabitants would require a vast store of energy to survive. That will present no problem, however, for by that time man will have learned how to obtain almost limitless amounts of energy from thermonuclear fusion and from such sophisticated technologies as the complete annihilation of matter and antimatter. It is also reasonable to assume that in time man will learn how to collect and to store most of the solar energy that now goes to waste in interstellar space. Thus, special energy-converting and -collecting installations will have to be designed and placed on such bodies as Mercury, Venus, and the moon, and over hundreds of millions or billions of years enough solar energy will have to be collected on these bodies to serve the needs of a wandering earth and its inhabitants for centuries.

What may appear to be a rather fanciful and farfetched scheme for the earth's escaping destruction when the sun becomes a red giant is not at all farfetched if one takes into account the vast technological developments, particularly in the realm of energy, that are bound to occur on the earth by that time. But even if the earth remains in its present orbit and is finally incinerated by an expanding sun, human life need not disappear, for vast colonizing expeditions can be sent out by mankind not only to the distant planets Jupiter, Saturn, Uranus, and Neptune but also to planets revolving around other stars. Quite recently scientists studied the cost of, and the problems involved in, sending the population of an entire town as large as Princeton, New Jersey, into space on a colonizing expedition, and they decided that to do so now would require one year's gross national product of the United States—about a trillion 1969 dollars. Such a vast expedition could be housed in a single space vehicle having the appropriate area and volume that would be powered and launched into space by hundreds of hydrogen bombs distributed over the surface of a 10-mile diameter hemispherical shield at the back of the vehicle. In other words, present-day nuclear technology is just about at a point where large-scale trips into outer space are conceivable but much too costly to be practical. Considering that nuclear technology is still in its swaddling clothes, one can easily accept the projection of man's capabilities outlined above and agree that regardless of what happens to the sun, humanity will survive.

12   The End of the Universe and the End of Time

The Death of the Sun

The sun must in time become a red giant because it will have transformed all the hydrogen in its core to helium and the gravitational energy released by the contracting pure-helium core will force the sun to expand. At the same time the temperature in the core will rise rapidly to over 100 million degrees, triggering a new series of thermonuclear reactions that will start with the transformation of helium to carbon and then proceed to the formation of the heavy nuclei, which will ultimately be fused into iron. How long the sun will remain in this heavy-nuclei-building state is not certain, but it is known that a very critical situation in the life of the sun will be reached when all the helium has been exhausted and the sun's interior consists of some carbon, oxygen, neon, and the like, but mostly of iron—say, 99 percent. The sun will be unable to maintain its gravitational equilibrium then because no nuclear energy will be released; gravitational collapse will necessarily occur, but what happens after that is not quite clear. It all depends very much on how violent the collapse is and on the "equation of state" of superdense matter, but this equation, which must be deduced from the general theory of relativity and nuclear physics, is still not fully determined. A number of different equations have been proposed that are similar in their overall features but that differ in certain details, depending on what one assumes about nuclear forces at very high densities. But even without discussing these fine points, one can deduce important overall characteristics that will help one to understand what will ultimately become of the sun.

Whether all the solar matter will be scattered into space to form a dispersed nebula after the gravitational collapse of the sun or whether most of the matter will settle down to form a white dwarf, a neutron star, or, ultimately, a black hole depends on what follows the gravitational collapse. If the collapse is so extremely violent that a vast amount of gravitational energy is released suddenly, the explosion that follows may, indeed, be so violent that most of the solar material will be propelled outward at many thousands of miles per second in the form of an expanding ionic envelope and observers living at that time on other planets in the Galaxy, and even in other galaxies, will see a supernova. What is of greatest interest is not the expanding ionic cloud that will give this phenomenon its appearance but the residual core, for depending on its mass and density, this core will either become a neutron star or a black hole. But the sun's becoming a white dwarf is the most probable eventuality.

White dwarfs are very small, but dense, white-hot stars with luminosities anywhere from 1,000 to 10,000 times smaller than the sun's; their densities range from a few hundred thousand grams to about 100 million grams per cubic centimeter. Enough of these stars have been discovered within a few light-years of the sun to indicate that white dwarfs are very numerous, as is to be expected if they are the dying stages of stars. A star will settle down to being a white dwarf only if its mass is less than 1.2 times the sun's mass; if the star's mass exceeds this value at the end of its orderly evolution, it will become a white dwarf only if it gets rid of some of its mass. If it cannot get rid of its excess mass, the star suffers gravitational instability and is forced into a catastrophic collapse that pushes it beyond the white-dwarf stage into a neutron-star or black-hole stage. Since the sun's mass is already below this critical value, the sun can become a white dwarf in an orderly manner after it has reached its iron phase. If the sun does not collapse too violently, it may go through a series of minor explosions following each other in intervals of a few hundred thousand years and thereby get rid of about 10 percent of its mass. It will then contract gravitationally in an orderly way until the iron nuclei are squeezed so closely together that the electrons will move around freely inside this iron lattice and, by their kinetic pressure, support the sun—now a white dwarf—against further collapse. The sun will exist in this state, called a degenerate electron state, for billions of years, and whatever planets are left will continue revolving around it pretty much the way they were before, but cold and dead.

If the gravitational collapse of the sun after its iron stage is extremely violent, the sun will become either a pulsar (a neutron star), spinning extremely rapidly and sending out intense bursts of electromagnetic radiation, or a black hole, a tiny superdense dead sphere of matter from which nothing, including light, can escape. Whether the sun will become a pulsar or a black hole after the sun's collapse will depend on how much the core is compressed by the collapse. If the compression of the sun's core is not too violent, so that its density does not exceed about 1,000 trillion grams per cubic centimeter the sun's core will become a neutron star, provided its mass is larger than about 5 percent of the sun's present mass. As the sun's core contracts during this violent implosion and as the core's density increases beyond white-dwarf densities—about 100 million grams per cubic centimeter—the free electrons that could support a white dwarf against collapse will be forced into the iron nuclei, where they will combine with protons to form neutrons. This will go on until the nuclei in the core contain too many neutrons to be stable. Neutrons will then drip off the nuclei in the core, slowly at first but more and more rapidly as the compression continues, until all the nuclei have disintegrated and only neutrons are left. Because such closely packed neutrons form a degenerate neutron gas, or a superfluid, and hence repel each other at this stage (densities of the order of 10 trillion grams per cubic centimeter), the solar core will be a stable neutron star with a 10-mile diameter. At first it will be pulsating very violently and rotating rapidly with a very strong embedded magnetic field, but the pulsations will die out very quickly, leaving a dense sphere that will spin on its axis about 30 times per second. As the very strong magnetic field of this final pulsar stage of the sun sweeps around during each rotation it will radiate strong electromagnetic waves. At this stage the surface of the solar pulsar will be a solid iron crust only a few centimeters thick but extremely dense.

If the implosion of the solar core is so violent and rapid that the nuclear repulsive forces cannot stop it, the collapse will slow down but not come to a complete halt at the neutron stage. The core density will continue to increase until the sun becomes a black hole, whose properties can be deduced from the general theory of relativity.

The general theory of relativity asserts that if the mass of the collapsing core of any star exceeds 3 solar masses, nothing can stop the core from contracting down to a black hole; the inward gravitational pull is so strong in such a case that the nucleon-nucleon repulsive forces are not strong enough to prevent the core from contracting right down to a point (in theory). Although the iron core of the imploding sun will certainly not meet this condition, since its mass will be smaller than that of the sun, it can still become a black hole if its density exceeds a certain critical value that is determined entirely by the mass of the core. To obtain this critical value of the density for a given mass, one divides the speed of light raised to the sixth power by 4.2 times the product of the cube of the gravitational constant and the square of the mass. We write this as

$$\text{critical density} = \frac{c^6}{4.2G^3M^2},$$

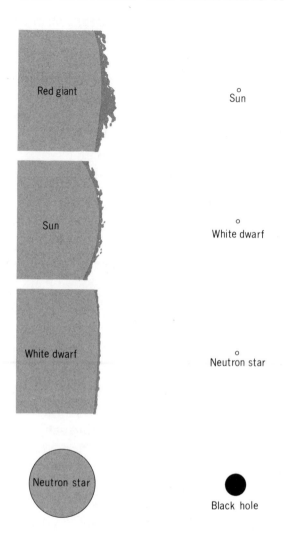

Comparative sizes of a red giant, a star like the sun, a white dwarf, a neutron star, and a black hole.

where c is the speed of light, G is the gravitational constant, and M is the mass of the core. This shows that the more massive a collapsing core is, the smaller its density need be for it to become a black hole. In fact, if a collapsing star had a mass equal to 100 million solar masses, it would become a black hole if its density were no more than that of water—1 gram per cubic centimeter. If after the initial solar implosion and the subsequent explosion of matter the collapsing core of the sun is not much smaller than the sun's present mass and if its density at any stage of its contraction exceeds 10,000 trillion grams per cubic centimeter, it will collapse down to a black hole. Its density will acquire the necessary excess if the collapse of the solar core is so fast that it is halted only momentarily, but not completely stopped, by the nucleon-nucleon repulsive forces of the neutrons when the density is slightly less than the figure given. The core will then rush past its neutron-star phase and contract further, thus forcing the density past the critical black-hole value. Everything will then be lost and no reversal or halting of the complete collapse will be possible; it will go on collapsing to a point.

One may ask what justifies the name *black hole* for such a collapsing object. Consider a spherical mass like the sun from whose surface particles are escaping; to do this, they must move with a speed equal to the speed of escape from this surface, which is determined entirely by the mass of the solar sphere and its radius. If the entire sphere is now squeezed down so that its radius is reduced by a factor of 4, everything on its surface becomes 16 times as heavy, the speed of escape from its surface becomes twice as large, and so on. Suppose now that this spherical mass contracts until the speed of escape from its surface equals the speed of light; this happens when the radius of the contracting sphere, called the Schwarzschild radius in honor of the German astronomer Karl Schwarzschild (1873–1916) who first solved Einstein's equations, equals $2GM/c^2$, where M is the mass of the sphere, G is the gravitational constant, and c is the speed of light. Since no light can escape from a sphere whose radius is smaller than this value, the sphere, on collapsing to this stage, would be completely invisible to any observer at a distance from it greater than the Schwarzschild radius. Not only can no light escape from the surface of such a sphere, but any light directed to the collapsed surface of this sphere from the outside can never be reflected out; it will be trapped in the curved space-time geometry of the sphere as soon as it comes closer to the surface of the sphere than the Schwarzschild radius. In other words, such a collapsed sphere can never be seen. Hence, the designation of this object as black. For this reason also the surface of the sphere whose radius equals the Schwarzschild radius is called the horizon of the black hole.

Now consider an astronaut in a spaceship falling freely toward the black hole; as long as he is outside the Schwarzschild radius, space and time have their usual properties, which means simply that the past and future do not depend on the direction of motion of the spaceship. He can move toward the black hole, move away from it, or move sideways as he pleases, and he can even stop himself; all he has to do is use the jet engines on his vehicle properly. But once he passes the Schwarzschild radius on his way inward, he no longer has any choice as to how he can move; he must move into the black hole no matter what he does or how much he tries to avoid it, ultimately becoming a part of the black hole itself. The reason for this is that in the geometrical region within the Schwarzschild radius, space and time are interchanged. This means that going from the past to the future in time is related to one, and only one, direction in space at a given point; that direction is in toward the black hole. An object can no more reverse its direction and go outward at that point than time can go from the future to the past. However powerful his jet engines may be and however much fuel he may use, the astronaut will never be able to reverse his inward fall once he has passed the Schwarzschild radius because falling into the black hole is equivalent to going from the past to the future. Thus, the name *black hole* is completely justified: all light passing the horizon of the black hole is absorbed by it and all objects penetrating the horizon fall into the hole.

Before leaving the black-hole death of the sun, consider

briefly how things would appear to an observer falling in with the sun as it collapses toward a black hole. Picture the observer as being completely protected against the sun's radiation by the proper kind of space suit, so that he cannot be vaporized by the intense heat. Initially, during the first stages of the sun's collapse, he would detect nothing unusual. Being in free fall like the sun's surface, he would see the sun's surface always at the same distance from him, and he would detect no gravitational field at all. He would, however, observe very dramatic changes on the sun's surface as well as in its rate of spin and in its magnetic field. As the sun contracted faster and faster, its magnetic field would increase greatly, it would rotate ever more rapidly, and the observer would find himself being dragged around ever more rapidly as the rotational speed of the core increased. By the time the solar core contracted to the neutron-star stage, the tidal action of the intense gravitational field of the neutron star would stretch the observer by a force of about 1 billion pounds and he would certainly not survive. He would have deduced long before the neutron stage that the sun was contracting rapidly and would probably realize what was in store for him if he continued falling in with the solar surface. Escape would still be possible, but if he did not escape at this point, he would collapse with the core down to the black-hole stage in a fraction of a second. At this point the radius of the core would be about 2.5 kilometers and the tidal forces would be so vast that they would tear the inward-falling observer to shreds. Even if he could escape being torn apart by the tidal action of the black hole, he could never reverse his precipitous descent into the hole, no matter how hard he tried, because of the relativistic properties of space and time already discussed.

Now consider how a distant observer at rest on a planet revolving around the collapsing solar core views the same set of events. Suppose that the planet Jupiter escapes destruction when the sun becomes a red giant and explodes, and suppose further that an astronomer on Jupiter is watching the core of

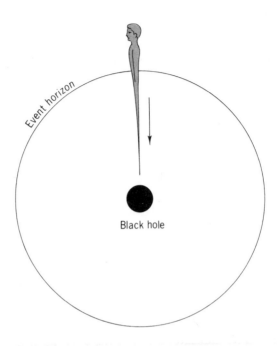

The descent of a person into a black hole. A person falling through the event horizon of a black hole would be stretched (and squeezed) by the tidal force into a thin line; once he penetrated the event horizon, no power in the universe could stay his fall or reverse his direction.

the sun contract down to its black-hole stage. He will detect the same general changes on the solar core's surface, in its magnetic field, and in its rotation as the freely falling ovserver does, but he will never see it actually become a black hole, because according to his clock it takes an infinite time for a contracting sphere to become a black hole. When the solar core begins to contract, the astronomer will see the surface of the core collapse catastrophically but then suddenly slow down as it approaches the Schwarzschild radius and finally appear to stand still as though it were frozen in time. However, it will get fainter and fainter very rapidly and suddenly, within a fraction of a second, fade from sight. The reason for this is that the amount of radiation that the collapsing core emits before it becomes a black hole is finite, so that the distant observer will find only a finite amount of light reaching him from the moment the surface of the core begins its catastrophic collapse to the moment it emits the last bit of radiation that can escape from it. Thus, the core will appear black to the distant observer after a finite lapse of time. There will be no change, however, in the orbit of Jupiter at that time or in any other phenomena in its neighborhood stemming from the gravitational field of the solar core. The collapse of the sun to a black hole will not affect the dynamics of the solar system, since the gravitational action of a spherical mass is always such that the mass appears to be concentrated at its center and hence independent of its size.

Tapping the Energy of Black Holes

One must distinguish between a nonrotating black hole, which is called dead because nothing can be done with it, and a rotating one, which is called living because energy can be extracted from it under proper circumstances. In the case of a rotating sphere that contracts down to a black hole two important and distinct distances from the center of the hole come into play. Expressed somewhat differently, there are associated with a rotating black hole two separate surfaces; one is called the event horizon, just as in the case of the nonrotating black hole, and the other, which does not exist for nonrotating black holes, is called the stationary limit, or frozen surface, which is always closer to the distant observer than the event horizon. This surface is called the stationary, or static, limit because no object within this surface can remain stationary, no matter how hard it tries. An astronaut with all the power in the world could not keep himself from moving if he were inside this limit. The frozen surface is the last surface of the collapsing core that a distant observer will see; space becomes timelike inside this surface if the black hole is not rotating. Objects and radiation can fall inward through the event horizon but can never emerge from it. But particles can emerge from the stationary limit if they have enough energy, and that leads to some interesting conclusions and a mechanism for extracting energy from a rotating black hole. Picture a particle penetrating the stationary limit and moving inward in the region, called the ergo-sphere, between the stationary limit and the event horizon of the rotating black hole. It can be shown from the equations of the general theory of relativity, when applied to rotating black holes, that a particle traveling through the ergo-sphere from the stationary limit to the event horizon can break up into two particles; one of these particles continues on its inevitable journey through the event horizon into the black hole, but the other one is ejected back out through the stationary surface to the distant observer with more energy than the initial particle had when it began its inward journey. This energy comes from the rotational energy of the black hole. This, then, would be a mechanism for extracting energy from the black hole the solar core would become if it became a black hole at all, and the amount of energy thus available would be extremely large—about 30 percent of the entire mass of the black hole could be transformed into energy by such a process. Of course, all of this energy would not be available in any one single process, but some of it would be released in each such process.

Using this idea, the American physicists Charles W. Misner, Kip S. Thorne, and John A. Wheeler, in their excellent treatise "Gravitation," have proposed a very ingenious but extremely fanciful scheme that a very technologically advanced civilization could use to extract a practically endless amount of energy from a rotating black hole. To see whether the Misner-Thorne-Wheeler technique can be applied to the earth some billions of years in the future, after the solar core has collapsed and is well on its way to becoming a rotating black hole, imagine that the earth and its inhabitants escaped destruction when the sun was expanding into its giant stage through the application of one of the technological feats outlined previously—the enlargement of the earth's orbit. With the solar core very nearly a black hole in this imaginative picture, it is spinning thousands of times per second, so that a great deal of rotational energy is available. Knowing all about the techniques for extracting energy from rotating black holes, scientists and engineers first construct a rigid framework around the solar core on which some vast cities are built. This framework will be placed at some convenient distance from the core but close enough to the stationary limit (perhaps about as far from the core as the moon is from earth now) to permit vehicles to be sent down through this stationary limit in a relatively short time. Suppose, now, that during the day trucks move through each city collecting the refuse and dumping it at special depots from which fleets of spacecrafts cart it down through the stationary limit. As each craft moves down toward this surface, the rapidly rotating gravitational field of the core drags the vehicle around in an orbit that spirals through the stationary limit and in toward the event horizon of the black hole. Shortly before the spacecraft reaches the event horizon, it dumps its load of refuse into the black hole, but does not penetrate the event horizon itself; in so doing it acquires more energy than it had when it was fully loaded at the beginning of its descent from the refuse depot. It is therefore ejected out of the ergo-sphere, through the stationary limit, and back to its starting point, arriving with much more energy than it started with. This excess kinetic energy is collected with special turbines that convert it into electrical energy. This process thus transforms refuse into electricity and the conversion is complete (except for the friction in the turbines). The amount of energy that can be obtained in this way is enormous, for it can be shown that when the refuse is dumped into the black hole all of its mass is converted into excess kinetic energy of the space vehicle in accordance with

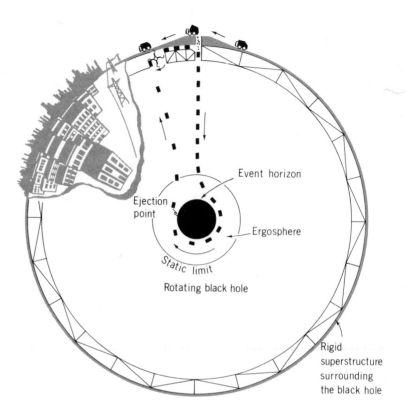

A city built on a superstructure surrounding a rotating black hole can extract energy from the hole by dumping its garbage into it.

Einstein's famous formula. In addition, some of the mass of the black hole is also converted into energy, which the ejected space vehicle also acquires.

Suppose now that each depot supplies a million tons of refuse and garbage to the black hole every day and that there are five such depots on the rigid structure. They then totally convert, via the black hole, more than five million tons of matter into energy every day; this is more energy than the present sun emits in one second in all directions and equal to the amount of energy the earth receives now from the sun in 67 years. There would therefore be no lack of energy for an advanced civilization orbiting the black-hole core of the sun. This may appear to be something in the nature of science fiction now, but there is no reason to suppose that the inhabitants of the earth in the very distant future will not use some such process to obtain the energy they need to survive.

Destruction from the Center of the Galaxy

Thus far the catastrophes discussed have been those that arise within the solar system itself, but it is clear that destructive events can occur outside the solar system also. Consider, for example, the possibility of some tremendous explosion at the center of the Galaxy that releases a vast amount of energy; it might set off a chain reaction in which millions of stars are annihilated in a very short time and whose energy thus released is great enough to destroy the earth. This may seem to be rather unrealistic because the earth is so far away from the center of the Galaxy (30,000 light-years) that any energy released there might be expected to be so greatly attenuated by the time it reached the earth that it would be hardly detectable. But consider the complete annihilation of a million stars, each as massive as the sun, at the center of the Galaxy. The energy released if all the mass of the million stars were transformed into energy would equal the amount of energy the sun would emit, radiating at its present rate, in 20 million trillion years; of this the earth would receive an amount of energy equal to the amount the earth now receives from the sun in ten years. This in itself would not be serious if the annihilation of the million stars at the center of the Galaxy occurred over a period of time longer than ten years. But it is conceivable that under certain circumstances, the annihilation of stars might occur very quickly—in a few months or even a few days. This could well be the case if the Galaxy collided with another galaxy consisting of antimatter. There is no observational evidence that such antimatter galaxies exist, but the laws of nature say that they should exist, so such possibilities must be considered. If the Galaxy did collide with an antimatter galaxy most of the annihilation of matter and antimatter would occur at the cores of the two galaxies because that is where the concentration of stars is greatest—about 30,000 times greater than in earth's part of the Galaxy.

If such a collision were to take place right now and the annihilation of stars occurred at the rate of two per minute, earth would in time receive from the center of the galaxy 10 times as much energy every second as it receives from the sun now. Inhabitants of the earth would receive no knowledge of this event during the next 30,000 years, the time it takes light to reach earth from the center of the Galaxy, but then a sudden flash would be seen, lighting up the sky with the brightness of 10 suns, and the earth would be enveloped in a searing flash of radiation. Even if only a fraction of the radiant flux reached the earth, it would probably be quite destructive of life because most of it would be in the form of X rays and cosmic rays.

Today there is strong theoretical and observational evidence of the collision of galaxies, particularly those that are members of a dense cluster of galaxies, such as the cluster in Virgo, about 40 million light-years away, which is known to contain at least 2,500 galaxies. In ordinary galactic collisions which have actually been photographed, the individual stars in the spiral arms of one of the galaxies do not collide with those in the other because these stars are too thinly scattered along the spiral arms, but stars in the cores of the two galaxies can collide. Unless one group of stars consists of matter and the other antimatter not much visible energy will be generated by the collision. However, the gaseous matter and the dust in the two galaxies will be set oscillating very violently and will thus emit large quantities of radio waves.

That a vast amount of energy can be released in the core of galaxies is shown by photographs of Seyfert galaxies (named after the astronomer who studied them), which appear to have exploding cores. Although the energies involved in these core explosions are large compared to typical stellar energies, they are far too small to do any damage to stars or planets in the arms of such galaxies.

The Heat Death of the Universe

While it appears that the earth is safe from galactic catastrophes, it is not safe from the various overall cosmological events that can, and ultimately will, bring things to an end. *An end* here does not mean that all matter will disappear but rather that a situation will occur where the orderly evolution and change that man sees going on all around him will cease. This will happen either because the universe has run down, like the spring of a watch, or because it has contracted down to a tiny, but highly concentrated, bit of matter. Consider first the running down of the universe as predicted by the second law of thermodynamics. Physicists found it necessary to introduce this law when they discovered that processes that occur naturally in any isolated system—a system that cannot exchange energy or matter with its surroundings—are irreversible and that these irreversible processes are such as to cause the system to progress from a state of order to a state of equilibrium—that is, a state of maximum disorder, or maximum homogeneity throughout the system. This law is formulated in terms of the entropy of the system: the greater the disorder, the larger the entropy. The second law is therefore equivalent to saying that in any isolated system the entropy can never decrease; the isolated system strives to attain a maximum value of its entropy and does so by evolving to states of maximum probability. The state of maximum entropy is thus a state of complete equilibrium, since such states are the most probable states.

Since the entire universe is itself an isolated system and therefore cannot exchange energy and matter with any outside system (by definition of the universe there can be nothing outside it), the second law of thermodynamics applies to it. This means that the universe must evolve in such a way as to maximize its entropy by changing from its present ordered, differentiated, inhomogeneous, and nonequilibrium state to a state of maximum disorder and maximum homogeneity. This final state of the universe—if it occurred—would be a state of equilibrium. That the present state of the universe is one of differentiation, order, and nonequilibrium is obvious from the existence of hot bodies (the stars) in a surrounding cold space; this means that the temperature of the universe is not the same at all points of space but changes discontinuously from a very low value in interstellar space to very high values in regions occupied by the stars. Owing to these temperature differences in various regions of space, there is a constant flow of energy from the hot stars to the cold surrounding space. Since the amount of nuclear energy in each star in the universe and in the matter in each cubic centimeter of space between the stars is finite, this unidirectional flow of energy will ultimately cease and the temperature of the universe will be the same everywhere, a few degrees higher than it is now between the galaxies. From that moment on, the universe will be cold, dead, and homogeneous, with no changes of any sort occurring anywhere; the entropy will be a maximum and complete equilibrium will prevail everywhere. In such a universe time will have no meaning, since nothing in the state of the matter then will differentiate between before and after. In isolated regions of space in such a dead universe there will be momentary fluctuations that will lead to a reduction of the entropy, with some inhomogeneity setting in, but these trends toward differentiation will quickly die out, so that the universe will never be able to attain any other state once it has reached this state of maximum entropy; this, then, will truly be "the end of the universe" if the universe does, indeed, follow the path toward complete uniformity. But there are forces that operate against this kind of "heat death" and promise a periodic rebirth of the universe; the observational evidence strongly points to a pulsating universe rather than to one that is running down.

Note that if the universe did lapse into entropy or heat death, it would not be when all the stars now in existence had used up all their nuclear fuel, for new stars are born from the material ejected from dying old stars, in which only a small part of the hydrogen present in the universe is transformed into iron. As long as the interstellar medium contains hydrogen and other elements lighter than iron, new stars like the sun can, and will, be formed, and they in turn will generate energy by transforming hydrogen and other elements into iron. In time, if nothing happens to prevent it, all interstellar matter and all the matter now in the stars will be transformed into iron through new star formation and all the nuclear energy available will be released. The universe will then, after many, many billions of years, if it happens at all, consist of black holes, dead neutron stars, dead white dwarfs, dead planets, and cold bits of matter floating around in a sea of cold electromagnetic radiation at a temperature of about 5° K.

The Collapse of the Universe

The final, all-inclusive catastrophe is the total collapse of the universe. All other catastrophes that occur will be blended together and merged in the ultimate collapse of the universe itself. The description of a star's collapse to a black hole provided a preview, on a very small scale, of the collapse of the universe itself, for the collapse of the universe resembles, in every respect except for size, the collapse of a star to a black hole.

Einstein's general relativistic equations, when applied to the universe, lead to three possible models. Two of these models are infinite (open) universes, which go on expanding forever until the density of matter in either of them is zero. If the universe were built according to either of these models, this too would be a kind of ultimate death, for such universes would in time be essentially empty—a finite amount of matter in an infinite space. For this reason these universes are not particularly interesting. The third model of the universe that stems from Einstein's theory is a finite (closed) expanding universe in which three-dimensional space is curved back on itself to form the three-dimensional surface of a hypersphere; the theory shows that such a closed universe cannot go on expanding forever but must stop and contract down to a compact structure and then expand again. This leads to a pulsating universe, which is extremely interesting because it promises a rebirth after the inevitable death that must follow the total collapse of the universe.

Until quite recently the observational evidence was not conclusive enough to permit one to choose the correct model of the universe from the three possible models, but now the evidence strongly favors the closed, pulsating universe. The observational work of Sandage, among others, shows that the expansion of the universe is slowing down at such a rate that in about 27 billion years it will come to a complete halt. By that time it may well be that all the stars now in existence and all those that are still to be born will have gone through all their evolutionary stages and practically all the matter in the universe will be either in the form of atoms of iron floating around in interstellar space or in the form of white dwarfs, neutron stars, and black holes. This state of the universe will be close to the heat death predicted by the second law of thermodynamics. But this quiescent phase will immediately be followed by a contraction. This will start very slowly but speed up as time goes on until it is as catastrophic as the last stages of the gravitational collapse of a star. The final stages of this universal collapse will occur about 40 billion years after the expansion has stopped, or about 67 billion years from now. When it does happen, it will sweep everything before it until all the matter and energy in the universe is again compressed by gravity into a tiny, compact kernel.

Some misconceptions about the appearance of distant objects when the universe stops expanding and begins to collapse must be corrected. When one looks out to the distant galaxies now, one sees them rushing away from the earth as though it were at the center of the expansion of the universe. One may therefore be left with the erroneous impression that these distant galaxies, having apparently—but only apparently—expanded farthest, will stop their recession from us first and begin their collapse first. But this is not so. The expansion of the universe is like the expanding surface of a toy balloon that is being blown up: just as the entire surface of the balloon stops expanding when one stops blowing it up, so too will all parts of the universe stop receding from each other when the expansion of the universe stops. And just as the surface of a balloon collapses as a single entity when the air is released, so too will the universe collapse as a single entity when expansion stops. The evidence for the change from expansion to collapse will come first to any observer who may be around from nearby galaxies. The Doppler shift of these galaxies, which is now toward the red, will first disappear and then become a blue shift. But when the nearby galaxies show a blue Doppler shift in their spectral lines, certain more distant galaxies will show no Doppler shift at all and galaxies at still greater distances will still show a red Doppler shift.

The reason for this spectrum of Doppler shifts is that the information carried to the future observers by the light from the distant galaxies will be old when it reaches them; it will tell them the way things were when the light first began its journey but not how things will be when it reaches them. Thus, if the universe has been collapsing for 10 million years and the observers are looking at galaxies that are farther away from them than 10 million light-years, the light from these galaxies will tell them how these galaxies were moving before the collapse began; in other words, the Doppler shift for this light will still be toward the red, and these galaxies, although actually approaching the observers, will appear to be receding.

As the galaxies approach each other during the collapse the gravitational force of attraction between them increases and the speed of the collapse increases; the potential energy of the gravitational field of the universe is gradually transformed into the kinetic energy of the galaxies as they rush toward each other with ever-increasing speed. At the same time, the overall temperature of the universe will begin to rise because the collapse will cause the interstellar atoms and molecules to move with ever-increasing speeds. Owing to the collisions between these interstellar particles, their initial motions of collapse will become random, which means that the temperature of space will rise. But even without the random motions of the interstellar molecules the temperature of the collapsing universe will rise because the wavelength of the overall background cosmic radiation will decrease; that is, the Doppler effect shifts the wavelength toward the blue. The basic laws of radiation show that the temperature of radiation in the universe rises as the wavelength of the radiation decreases.

Eventually the collapse of the universe will bring the stars that are not black holes in the galaxies so close together that collisions between them will tear them apart and the stellar matter they contain will be smeared out into a uniform gas. If black holes in the universe collide, they will merge to form larger black holes, and if a star collides with a black hole, the matter in the star will be swallowed up by the black hole. By this time the temperature of the universe will have risen to millions of degrees and the radiation itself will disrupt any stars that are still around. But, in time the radiation, too, will be swallowed up by black holes and then the black holes themselves will begin to merge until a single black hole is left, which will be the final state of the universe. Shortly before this happens, the universe will resemble the fireball from which it began, but now it will be an imploding and not an expanding fireball.

The Rebirth of the Universe

All the matter in the universe must ultimately collapse to a single black hole, but what will happen after that? The general theory of relativity does not give a definite answer, nor is there any way to carry out any kind of experiment on black holes that will reveal what will ultimately happen to a black hole. The reason for this is that as long as one is outside the black hole, one can never see anything below the event horizon of the black hole; although the black hole can be safely observed from the outside, no information can be gained about events within the black hole itself. And if any probe were sent down through the event horizon, it could never return to tell one what the inside of a black hole is like. Information about the interior of a black hole must come from a correct theory, just as man's knowledge of the evolution of stars does. But the only two theories man has at the present time are the general theory of relativity and the quantum theory. Unfortunately, however, no one has yet forged these two into a single quantized general relativistic theory that will reveal what happens when the collapse continues down to the most minute dimensions. The general theory of relativity says that nothing can stop the collapse, because after the Schwarzschild radius is reached, the direction in time from the past to the future is necessarily tied to the continuing collapse down to a point. But when the universe gets down to very small dimensions and one considers events occurring in very tiny volumes of space, the quantum theory cannot be neglected. It is believed today that the ultimate collapse of the universe to a point will not happen, because quantum effects will come into play and prevent ultimate collapse.

Here is a situation similar to the one that physicists faced before the Bohr theory of the atom was introduced around 1913. Everything known at that time about the interaction of electric charges and the motion of a particle acted on by a force led to the inescapable conclusion that matter as man knows it should not exist, because all atoms ought to collapse under the mutual attraction of the positively charged nucleus of the atom and the negatively charged electrons. Quantum theory provides the answer: The existence of a quantum of action places very definite restrictions on the smallness of a region to which a particle having a given speed can be confined. The smaller the region of confinement, the faster the particle moves until its kinetic energy becomes so great that it tears through all barriers and escapes confinement. An atom cannot collapse, because the electrons would move so violently on being brought close to the nucleus that they would simply fly out. If the same ideas are now applied to the collapsing universe, one can argue that as the implosion of the universe continues into the black-hole stage and beyond, the quantum fluctuations described above will impart such violent motions to the particles in the black hole that some kind of violent rebound or explosion will again occur and that another expansion phase of the universe will begin, leading ultimately to another cycle of star-building, synthesis of elements, and, finally, planets and life itself.

The exact details of how the new expanding universe will emerge from the black hole of the old universe are not clear, since there has not yet been a complete unification of the general theory of relativity and the quantum theory, which is needed for a detailed understanding of how the imploding universe that has become a black hole is transformed into an expanding one.

Without a complete quantum–general-relativistic theory, one can only make some intelligent guesses about the imploding universe, based on present knowledge of the way electrical systems behave when the quantum theory is taken into account. The quantum theory of charged particles reveals that an electron directed toward a proton does not collapse onto the proton when it gets very close to the proton, as one would expect from classical Newtonian physics, but, instead, is thrown out in a random direction, just as a rubber ball bounces randomly off an irregular surface. Now the essence of the quantum theory, as applied to the scattering of electrons by the proton, is that it does not permit one to describe the interaction of an electron and proton in detail and, thus, to predict exactly what direction the electron will take; it permits one only to calculate the probability that the electron will move in some particular direction. This uncertainty in the behavior of the electron is the mark of the quantum theory. If one now applies the same general quantum ideas to the collapse of the universe, one may conclude that the imploding bits of the universe will not go on collapsing to a point but will be scattered into a new expansion cycle. But just as one cannot predict which direction an electron will take when it bounces off the proton, so one cannot predict what the new expansion cycle of the universe will be like. If everything were completely determined, as in Newtonian theory, all future expansions of the universe would be identical and every event that occurred, and will occur, in the present universe would occur again and again in precisely the same spatial and temporal order in future universes. But the quantum theory is not a completely deterministic theory; it can only give probabilities for this or that kind of universe. This means that the same general pattern of stars, galaxies, planets, and living organisms will occur over and over again as one expansion cycle follows another, but no single event or object—except the basic particles like electrons, protons, neutrons, atoms, and molecules—in one expansion cycle will be the same as any single event in any other cycle. There will be stars in the next cycle, but no star just like the sun; planets, but no planet like earth; and living creatures, but none just like you.

One may compare all the future universes just before

they begin their expansions to the millions of fertilized eggs that chickens lay in a year: the yolks and whites of all these eggs look identical, and yet no two chickens hatched from these eggs will be the same, because the genetic codes programmed in the DNA molecules of these eggs are all different. One can predict that each egg will evolve into a chick, but not the characteristics of any single chick. In the same way, the compact kernel of matter into which the universe will collapse in any one cycle is programmed to become another expanding universe, with all the features that the present expanding universe has, but it will be as unlike the present universe as one chick is unlike another. In connection with this, note that the code that is programmed into the compact state of the universe before the big bang occurs is quite different from the genetic code in a fertilized egg; whereas the latter is a structural code involving a variety of patterns, the former is a code of natural laws, forces, and basic constants—for example, the gravitational constant, the speed of light, and Planck's constant of action. The code in the preexpansion state of the universe is the most fantastic and amazing example of coding imaginable, for it includes all other codes that nature or man can devise.

Inevitably one must ask if the laws and constants of nature are the same for all time, past and future, or if they change in an unpredictable way each time the universe begins a new 80-billion-year expansion phase? Wheeler has argued very cogently and strongly in favor of a random change not only in the constants of nature but also in the very laws themselves. His argument runs somewhat as follows: Just as an event in the present universe can be represented as a point in four-dimensional space-time, so the whole universe must at any moment be represented by a point in a superspace with an infinite—or, at least, an extremely large—number of dimensions or coordinates instead of just the four dimensions of space-time. The coordinates of a point in this superspace give the location of every event in the universe, and as the universe evolves, this point in superspace moves about indicating the change in the size of the universe, the positions and motions of all the particles in the universe, and the curvature of space-time. The mathematical description of the motion of the point in superspace that represents the evolution of the universe must take into account the quantum fluctuations in space-time at all points in the universe, just as the description of the motion of an electron in any electric field in the universe must take account of quantum effects at points occupied by the electron. This is done for the electron by assigning to it in space-time a Schrödinger wave function, which gives the probability for the electron's being at a certain point and having a certain velocity at a certain time. In other words, the quantum fluctuations force us to describe the electron by an array of probability amplitudes in space-time. This is why one cannot predict precisely how the electron will move after it is scattered by a proton.

From this Wheeler argues that the motion of the point in superspace that represents the evolution of the universe must be described by a wave function or a set of probability amplitudes because of quantum fluctuations in superspace. As the universe evolves following the initial explosion these fluctuations have a very small—in fact, negligible—effect on the history of the universe, but these very fluctuations become increasingly more important during the final contraction phase of the universe. Finally, when the universe has collapsed down to a radius of about one-hundredth of a trillionth of a centimeter, the quantum fluctuations become so violent that a new expansion phase begins. But owing to these very quantum fluctuations, there is no way, according to Wheeler, to predict the trajectory of the "reprocessed universe" in superspace, since the new point must be represented by an array of probability amplitudes. This means that the reprocessed universe would be governed by a whole new set of numerical values for the natural constants. If this were so, the chance for life during most expansion-contraction cycles of the universe would be extremely small, since only a small range of values of the constants of nature is compatible with life; the duration of an expansion-contraction cycle could be as short as a few million years or as long as a few trillion years, depending on just which unpredictable numerical values for the constants happen to be yielded by a future big bang.

There are fatal flaws in this argument that stem from the very base on which it appears to rest—namely, the quantum theory or its equivalent, the uncertainty principle. The quantum theory and all its consequences are the result of Planck's constant of action h, which is very small (6.7 divided by 1,000 trillion trillion erg seconds). Such things as the uncertainty principle, the scattering of electrons by protons, the structure of atoms, the unpredictability of the exact path of an electron, and the unpredictability of the precise evolutionary stages of the universe are consequences of this remarkable constant. If the value of h were zero, there would be no uncertainty principle and no unpredictability. Wheeler's entire argument rests on the existence of the constant of action h, from which he concludes that the constants of nature change in an unpredictable way every time the universe is reprocessed after its collapse. But this can be so only if the value of h itself changes in an unpredictable way with each new big bang. This leads to a kind of circular reasoning about the numerical value of the constant h itself. If Wheeler is right, the value of h in the present universe must in some manner or other lead to fluctuations in the value of h itself and hence to an unpredictable value of h in future universes. Note that only if the value of h changes will the value of the electric charge on the electron and the masses of the elementary particles change, because they themselves are quantized quantities. But the value of h cannot itself change at the very time that it is dictating the probability distribution of the values of all other constants.

If the value of h did change in a random way, the existence today of a nonzero value of h would be difficult to understand if one assumed that an infinite number—or, at least, a very large number—of expansion-contraction cycles occurred in the past. The theory of probability asserts that if the value of h fluctuates in a completely random way, with x being the average single change in its value, then the value of this constant after n cycles of expansion and contraction will differ from h by the amount $\sqrt{n}\,x$. Since positive changes of h would be as probable as negative changes, it is clear that h must then have equaled zero, or very nearly so, an infinite number of times in the infinite past—or a very large number of times in the last trillion billion years. But this leads to a contradiction, for if h had ever been zero in the past, no further fluctuations in nature could have occurred and all expansions following that event would have been precisely determined. Moreover, no quantum effects could exist in subsequent universes nor could any atoms, stars, planets, or living beings. Man's very existence disproves the possibility of random fluctuations in h from one expansion-contraction cycle to the next and hence in the constants of nature; indeed, man's existence implies that life will occur over and over again, but not precisely as it evolved in the present universe, for the normal fluctuations that occur in all physical systems will change the initial conditions of each new expansion phase of the universe, so that no two such phases will be identical. Thus, men have (in their own existence) not only the promise of life renewed but also the promise of almost infinite variety in such life.

Epilogue

The story of the birth and death of the world told here spans some 80 billion years, showing how it all began in chaos some 13 billion years ago and must all end in chaos some 67 billion years hence. As one contemplates this vast cyclic panorama, in which a world of order, symmetry, and intelligent life arose from disorder, as the culmination of a series of natural events, one is struck by the incredible contrast, bordering on what appears to be incompatibility and antagonism, between one realm of nature, life, and the antagonistic realm of the inanimate universe surrounding it. Every step in the formation of the complex organic molecules that constitute a living organism can be explained in terms of natural laws and the basic electromagnetic forces that govern all chemistry, but an understanding of the essence of life eludes us; therein lies the apparent contradiction between the living and the nonliving. Life is not simply the sum of all the molecules and atoms in a living creature at any one moment and the forces that keep these particles together, as is a stone, but rather a complex, unchanging dynamical pattern—a kind of molecular dance, with very precise and well-laid-out steps—that maintains itself from generation to generation. The sameness of the species in each generation and of every single living being from day to day is preserved, even though the molecules in every living organism are constantly being replaced by new ones of the same kind, which take their proper places in this molecular dance of life. Now the overwhelming mystery and apparent contradiction presented

318

by these everlasting and precise dynamical patterns, without which life would be impossible, is that they coexist with—indeed, that they give rise to—patterns in living creatures that appear to be imprecise, unpredictable, capricious, and without any fixed structure. The reference here is to what is usually called free will—the freedom of choice that every human has to perform or not to perform a variety of apparently unrelated acts. All the molecules in one's body behave in accordance with the laws of the quantum theory in well-defined ways that, although not strictly deterministic, owing to the quantum principle of indeterminacy, are statistically precise. Yet one is not governed by this precision, for one can move his muscles as he pleases, think what he pleases, and respond to sensations in a variety of ways. How can such macroscopic unpredictability stem from microscopic predictability? Since the fluctuations arising from quantum indeterminacy can be shown to be negligible when large masses, such as the human brain, are concerned, one cannot ascribe the acts of free will to quantum indeterminacy; they must stem from a qualitative change in the behavior of molecules when they form large aggregates.

But these various random and unpredictable acts that all human beings perform from moment to moment are actually less capricious than they appear, for although thoughts of all sorts flit through people's minds constantly and all kinds of minor things are done by people in a random way, they all add up to overall patterns that are not only predictable but common to all men. Moreover, these overall predictable behavioral patterns distinguish man from all other living species, each of which is marked by its own set of behavioral patterns, which also emerge from a sea of random acts. However randomly one idea follows upon another in the thoughts of each person and however arbitrary and unpredictable the momentary actions of each may be, there is a quality common to all that people do and think that permits each to relate his or her individual consciousness to that of everyone else. Without this commonality or sense of identity, communication between individuals would be impossible and life would cease.

In recognizing the existence of these overall patterns, one becomes aware of two distinct groups of such patterns; one of these groups is essential not only to the survival of the individual but to life itself, but the other, having no apparent survival value, presents the greatest mystery of all. The first group includes all those responses to the environment that are necessary for the protection and sustenance of life, such as the avoidance of danger, the escape from pain, and the satisfaction of hunger and sexual desires. It is easy to understand and to classify such drives in terms of survival value, and one can see how they fit into nature's plan for the propagation of life, but these survival responses and the incredible clinging to life that are common to all living beings do not explain life or justify its existence. But an explanation of, and reason for, life may be found in the second group of patterns, which are most pronounced and highly developed in mankind but are probably present on a very much smaller scale in lower animals: all those activities and human urges that are associated with truth, love, beauty, justice, charity, mercy, a compassion for living things in general, and other virtues that are often pursued and practiced at a loss to one's own well-being. Here, also, in these purely cerebral activities, one perceives patterns common to all human beings; that great musicians, artists, scientists, poets, and philosophers can communicate their ideas to others and generate in others similar thoughts establishes that universality of all minds which coexists with individual identities.

Of all of these virtues, the search for truth is the most remarkable and is, perhaps, the reason for intelligent life itself, as though nature were seeking to satisfy its own intellectual curiosity by examining and comprehending the universe through the eyes of a thinking being. *Truth* here does not mean the truth thought of in one's daily activities, where it is essential to know the "truth" about the time, the weather, one's health, or one's financial state. The virtue of such pragmatic "truths" is obvious, for without them the constant uncertainty about routine events would make life intolerable. Rather, it means that truth expressed in the laws that govern the universe and reveal to man the universal properties of space, time, and matter. These properties are often so obscure and hidden from direct view that they cannot be discerned in isolated facts or deduced from unrelated events but must be derived from fundamental laws. Now the curious thing about this aspect of truth is the remarkable appeal that it has for the human mind; and its pursuit, which motivates man more powerfully than anything else, clearly distinguishes man from the rest of the animal kingdom. This pursuit is the most glorious and exalting experience of which man is capable, and such truth possesses elements of beauty. Emily Dickinson speaks of "one who died for beauty" as being a brother to "one who died for truth," since the "two are one." And Edna St. Vincent Millay opens her sonnet on Euclid with the line "Euclid alone has looked on beauty bare," thus equating the truth of euclidean geometry with the purest form of beauty. Again, Shakespeare, in one of his sonnets, notes that truth enhances beauty:

> O, how much more doth beauty beauteous seem
> By that sweet ornament which truth doth give.

Man also assigns a quality of truth to the beautiful and tends to equate any intellectual or emotional experience that exalts him to a revelation of some great truth. Men thus speak of a "valid" experience or a "moment of truth" or the "validity of a religious revelation." But this exaltation of emotion and sensation to the level of great truth obscures the nature of scientific truth and opens the door to mysticism and metaphysics, which have no place in science.

That such drives toward truth and beauty exist and stem from the same molecular patterns that drive man to satisfy his hunger is the greatest of all mysteries unless one supposes that these very mysteries are not to be explained by all the events that preceded the emergence of life but rather that these very mysteries, which are the most sublime manifestations of life, explain and give reason for everything else that went before. One would then have to say that the big bang, the expansion of the primordial matter stemming from it, the formation of the galaxies and the oldest stars, the nucleosynthesis of heavy elements from hydrogen and helium in the hot interiors of these stars, the nova outbursts on the death of these stars, and the consequent formation of planetary systems like earth's were all in preparation for the greatest event of all—the appearance of intelligent life.

But one may ask why, if intelligent life is the reason for the universe, was it necessary for such vast and complex cosmic preparations before life emerged? Scientists often say that nature "solves all its equations at once and exactly" and that everything moves and behaves in accordance with these equations. Why, then, having the solutions of all its equations at hand, and presumably knowing all the end results, could not nature have set the stage for a cosmic genesis and created a complete intelligent being in a single act? The reason lies in the quality of the natural laws themselves and in the mathematical properties of the equations; the laws are quantum laws and the equations are differential equations. Atoms and molecules can exist only because action is quantized; but this very important feature of the universe, which is essential for the existence of the great variety of matter, introduces an ineluctable indeterminacy (the famous Heisenberg principle, which states that the position and motion of a particle—for example, of an electron—cannot both be known simultaneously with infinite accuracy) that prevents precise knowledge of future events, even if the solutions of all the equations that govern all events are known. The reason for this is that solutions of differential equations can predict the future only if the present is precisely known;

but quantum indeterminacy, which is an essential part of nature, prevents this. Now this indeterminacy is not only a restriction on man's ability to make precise measurements; it is a restriction built into the laws of nature themselves, so that there can be no infinite intelligence that can know all things and predict all future events. This means that nature, with all its wisdom, cannot construct a perfect being in a single act because quantum indeterminacy makes the knowledge of what is perfect unattainable even to an infinite intelligence if there were such an entity in the universe.

The development of beings of ever-increasing intelligence must therefore occur in that piecemeal manner called evolution, with the quantum fluctuations at each stage of development determining the new forms that can arise from the old ones. It is customary at this point to invoke such concepts from the theory of evolution as random mutations, natural selection, and survival values to explain the emergence of complex living forms from simple forms; but such concepts cannot possibly explain many highly developed and extremely complex attributes and characteristics of advanced living organisms that neither have survival value now for the organisms nor had any such value at any stage of their development. Even where some particular constituent of an organism, such as the venom of a snake, has obvious survival value, random mutations and natural selection cannot explain its final emergence from earlier stages in its development when it had no survival value, when it was a nonpoisonous protein. How could random mutations and natural selection have forged the simple inoffensive proteins of a nonpoisonous presnake organism into the virulent venom of a rattlesnake or a cobra without some kind of prescience about the end product? If one accepts such prescience, one is thrown into the realm of mysticism, which is contrary to the whole concept of evolution; but if one does not accept this, one cannot understand, using the present concepts of evolution, how there can be any direction in the evolution of a constituent of an organism during periods when the constituent has no survival value.

There is a way out of this difficulty that not only explains the evolution of biological components, such as venomous proteins from nonpoisonous proteins, but also the development of the intellectual attributes of human beings. Random mutations and natural selection play an important role in evolution but by no means the dominant role, which must be assigned to what might be called the potential for change or evolution in a certain direction. It may be that programmed into each gene or perhaps each DNA molecule in the nucleus of a germ cell is a potential for change in a certain preferred direction that depends on the structure and spatial symmetry of the arrangements of the nucleotides in that molecule; the greater this potential, the greater is the probability that an evolutionary change in the preferred direction (which is programmed into the molecule itself as the result of the changes up to that point) will take place. As evolutionary changes in some particular gene occur, the symmetry, and hence the potential for additional change, in that gene may be enhanced or diminished; but as time goes on, a point of saturation is approached, and changes become less and less frequent until all significant changes stop. Thus, one may compare the vast evolutionary proliferation of living organisms to a huge maze with many bypaths that lead to dead ends but with one main, open-ended track, along which mankind is progressing.

The evolutionary process suggested here is similar to the process that leads from simple atoms to complex molecules in a given mixture of various kinds of atoms and molecules under appropriate conditions. As noted, a hierarchy of complex molecules evolves quite naturally out of interacting atoms if they have enough energy and are allowed to come into contact with each other. An examination of such a mixture reveals a whole range of molecules, some of which undergo no changes at all while others change into more complex molecules; each molecule that is formed either carries with it the potential for combining with other molecules and becoming more complex or has exhausted that potential and proceeds no farther. Applying this idea on a much vaster scale, to the evolution of living organisms, one sees that the direction in which an organism will evolve is dictated by its previous history and development, and among all such organisms there will always be one species—the main track in the maze—in which the probability for evolution to higher and higher forms will be a maximum. Man is this species on the earth, and his development toward perfection, a state that cannot be defined at any stage in the development of man, is inevitable; this is entirely a consequence of the symmetry and three-dimensional structure of his DNA molecules, which in turn stem from the basic forces in nature, and which at each stage are a product of the changes that were dictated by the symmetry of the molecules at a previous stage of their development. Thus, the very chance and randomness that prevent man from predicting the future are the tools that nature uses to insure not only the emergence of life in each expansion cycle of the universe but also the approach to perfection and complete harmony as more and more complex forms of life evolve.

CONVERSION TABLE

BIBLIOGRAPHY

INDEX

Conversion Table

LENGTH

metric	U.S. equivalent	U.S.	metric equivalent
centimeter	0.394 inch	inch	2.54 centimeters
meter	39.370 inches 3.281 feet	foot	30.5 centimeters 0.3048 meter
kilometer	0.621 mile	mile	1.609 kilometers

AREA

square kilometer	0.385 square mile	square mile	2.590 square kilometers

WEIGHT

gram	0.035 ounce 0.002 pound	ounce	28.350 grams
kilogram	2.204 pounds	pound	453.592 grams 0.453 kilogram
metric ton (1000 kilograms)	2204.000 pounds	ton, short (2000 pounds)	0.907 metric ton

Bibliography

Abetti, Giovanni. *Exploration of the Universe.* London: Faber & Faber, 1968.

Aller, L. H. *Atoms, Stars and Nebulae.* Cambridge, Mass.: Harvard University Press, 1971.

Baade, Walter. "Stellar Populations." In *Evolution of Stars and Galaxies,* ed. Cecilia Payne-Gaposchkin. Cambridge, Mass.: Harvard University Press, 1963.

Bergamini, David. *The Universe.* New York: Time-Life Books, 1962.

Bok, B. J., and Bok, P. F. *The Milky Way.* Cambridge, Mass.: Harvard University Press, 1957.

Bondi, Hermann. *Cosmology.* Cambridge: Cambridge University Press, 1960.

————. *Rival Theories of Cosmology.* Oxford: Oxford University Press, 1960.

Burbidge, Geoffrey, and Burbidge, Margaret. *Quasi-Stellar Objects.* San Francisco: Freeman, 1967.

Dicke, R. H. *Gravitation and the Universe.* Philadelphia: American Philosophical Society, 1970.

Fowler, W. A. *Nuclear Astrophysics.* Philadelphia: American Philosophical Society, 1967.

Gamow, George. *The Creation of the Universe.* New York: Viking, 1952.

Glasby, J. S. *Boundaries of the Universe.* Cambridge, Mass.: Harvard University Press, 1971.

Hawkins, Gerald S. *Splendor in the Sky.* New York: Harper & Row, 1969.

Hoyle, Fred. *Galaxies, Nuclei and Quasars.* New York: Harper & Row, 1965.

————. *The Nature of the Universe.* Oxford: Oxford University Press, 1960.

————. *Frontiers of Astronomy.* New York: Harper, 1961.

Hubble, Edwin P. *The Realm of the Nebulae.* New York: Dover, 1958.

Kopal, Zdenek. *Man and His Universe.* New York: Morrow, 1972.

Lovell, A. C. B. *Exploration of Outer Space.* New York: Harper, 1962.

331

Lovell, A. C. B. *Our Present Knowledge of the Universe.* Cambridge, Mass.: Harvard University Press, 1967.

Lovell, A. C. B., *et al. The New Universe.* New York: Giniger–Rand McNally, 1968. See especially articles by R. H. Dicke, J. L. Greenstein, Frederick Reines, and Maarten Schmidt.

Menzel, Donald, Whipple, F. L., and Vaucouleurs, Gerard de. *Survey of the Universe.* New York: Prentice-Hall, 1970.

Misner, C. W., Thorne, K. S., and Wheeler, J. A. *Gravitation.* San Francisco: Freeman, 1973.

Motz, Lloyd. *This Is Outer Space.* New York: New American Library, 1962.

———. *This Is Astronomy.* New York: Columbia University Press, 1963.

———. *Astrophysics and Stellar Structure.* Waltham, Mass.: Ginn-Xerox, 1970.

———. (ed.). *Astronomy A to Z.* New York: Grosset and Dunlap, 1964.

Motz, Lloyd, and Duveen, Anneta. *Essentials of Astronomy.* New York: Columbia University Press, 1971.

Russell, H. N. *The Solar System and Its Origin.* New York: Macmillan, 1935.

Shapley, Harlow. *Galaxies.* Cambridge, Mass.: Harvard University Press, 1961.

Singh, Yagit. *Great Ideas and Theories in Modern Cosmology.* New York: Dover, 1961.

Sitter, Willem de. *Kosmos.* Cambridge, Mass.: Harvard University Press, 1951.

Urey, H. C. *The Planets, Their Origin and Development.* New Haven, Conn.: Yale University Press, 1952.

Whipple, F. L. *Earth, Moon and Planets.* Cambridge, Mass.: Harvard University Press, 1963.

Woltjer, Lodewijk (ed.). *Galaxies and the Universe.* New York: Columbia University Press, 1968.

Young, Louise B. (ed.). *Exploring the Universe.* New York: McGraw-Hill, 1963.

———. (ed.). *The Mystery of Matter.* New York: Oxford University Press, 1965.

Index

About the Author

Lloyd Motz is professor of astronomy at Columbia University, New York. He has done extensive work in stellar interiors, nuclear physics, thermonuclear reactions, unified field theory, geometrical optics, and in other areas of science. Dr. Motz has received many awards; in 1960 he was the first recipient of the prize of the Gravity Research Foundation, for a new theory of the structure of fundamental particles; and in 1972 he received the Boris Pregel Award of the New York Academy of Sciences for his contributions to physics and astronomy. He has published numerous scientific papers and several books, including *This Is Astronomy* and *This Is Outer Space*. Most recently he was co-author of *World of the Atom*.